John Muter, South London School of Pharmacy, South London Central Public Laboratory

A Manual of Analytical Chemistry

John Muter, South London School of Pharmacy, South London Central Public Laboratory

A Manual of Analytical Chemistry

ISBN/EAN: 9783742862785

Manufactured in Europe, USA, Canada, Australia, Japa

Cover: Foto ©berggeist007 / pixelio.de

Manufactured and distributed by brebook publishing software (www.brebook.com)

John Muter, South London School of Pharmacy, South London
Central Public Laboratory

A Manual of Analytical Chemistry

A MANUAL
OF
ANALYTICAL CHEMISTRY,

QUALITATIVE AND QUANTITATIVE— INORGANIC AND ORGANIC.

ARRANGED ON
THE PRINCIPLE OF THE COURSE OF INSTRUCTION GIVEN AT
THE
SOUTH LONDON CENTRAL PUBLIC LABORATORY,
AND THE
SOUTH LONDON SCHOOL OF PHARMACY.

BY

JOHN MUTER, M.A., Ph.D., F.R.S.E., F.I.C., F.C.S.,

DIRECTOR OF THE LABORATORY; ANALYST TO THE METROPOLITAN ASYLUMS BOARD; PUBLIC ANALYST FOR LAMBETH, WANDSWORTH, SOUTHWARK, NEWINGTON, BERMONDSEY, ROTHERHITHE, AND TENTERDEN; PAST PRESIDENT OF THE SOCIETY OF PUBLIC ANALYSTS; AUTHOR OF "PHARMACEUTICAL AND MEDICAL CHEMISTRY" (THEORETICAL AND DESCRIPTIVE), "A KEY TO ORGANIC MATERIA MEDICA;" EDITOR OF "THE ANALYST," ETC., ETC.

THIRD EDITION, ENLARGED, WITH ILLUSTRATIONS.

PHILADELPHIA:
P. BLAKISTON, SON & CO.,
1012 WALNUT STREET,
1887.

PREFACE.

In the present edition of this book a condensation in bulk has been attained, but, by help of a change in the style of printing, a greatly increased amount of special matter has at the same time been introduced. In writing such a book it is always difficult to draw the line as to the exact extent to which it should go, but I trust that I have fairly succeeded in adopting a wise limit.

The object of the work has been to produce a manual at once short and easily understood, but yet taking the student from the simplest to the most complex matters of qualitative analysis, and also dealing with quantitative work sufficiently to give him a fair insight into all branches of this department. Students preparing for special examinations will find in the book information suitable for every Pharmaceutical or Medical and most general University Examinations in Practical Chemistry now in existence; while the student who works through it will become a sufficiently expert general analyst to enable him to decide to what speciality, and consequently to which of the large standard works, he will devote his future attention. Teachers will also, I hope, find the present edition specially worthy of attention, as it is divested of those minute details which can really only be supplied by personal supervision. Every process given is the result of twenty years' experience in the teaching of practical chemistry, and all have been carefully tested by actual work in the hands of students, except where official processes of the Pharmacopœia are necessarily introduced. The acid course is, so far as I know, the only one published that goes on like the ordinary metal course, and certainly and successively detects all the acids in one direct line of experiments.

In conclusion, I trust that an equal amount of favor will be shown to the present edition as has been extended to the past ones, both in Great Britain and in America.

J. M.

WINCHESTER HOUSE, KENNINGTON, LONDON, S.E.,
September 1887.

TABLE OF CONTENTS.

PART I.—QUALITATIVE ANALYSIS.

CHAPTER I.

The Processes Employed by Practical Chemists.

	PAGE
1. Solution	1
2. Lixiviation and Extraction	2
3. Precipitation	2
4. Decantation	3
5. Filtration	3
6. Distillation	4
7. Sublimation	4
8. Fusion	4
9. Evaporation	5
10. Crystallisation and Dialysis	5
11. Electrolysis	6
12. Pyrology	7
13. Preparation of sulphuretted hydrogen	9

CHAPTER II.

Detection and Separation of the Metals.

	PAGE
Group Reagents	10–11
Group I.	11–13
1. Silver	11
2. Mercurosum	12
3. Lead	12
Group II.	13–18
Division A.	13–15
1. Mercuricum	13
2. Bismuth	14
3. Copper	14
4. Cadmium	15
Division B.	15–18
1. Arsenic	15
2. Antimony	16
3. Tin	17
4. Gold	17
5. Platinum	18
Group III.	18–24
Division A.	18–21
1. Iron	18
2. Cerium	20
3. Aluminium	20
4. Chromium	21
Division B.	21–24
1. Manganese	21
2. Zinc	22
3. Nickel	23
4. Cobalt	23
Group IV.	24–26
1. Barium	24
2. Strontium	25
3. Calcium	25
Group V.	26–28
1. Magnesium	26
2. Lithium	26
3. Potassium	27
4. Sodium	27
5. Ammonium	27

CHAPTER III.

Detection and Separation of Acidulous Radicals.

	PAGE
1. Hydrofluoric Acid and Fluorides	29
2. Chlorine, Hydrochloric Acid, and Chlorides	29
3. Hypochlorites	30
4. Chlorates	30
5. Separation of Chlorates and Chlorides	30
6. Perchlorates	30
7. Bromine, Hydrobromic Acid, and Bromides	30
8. Detection of Chlorides in the presence of Bromides	31
9. Hypobromites	31
10. Bromates	31
11. Iodine, Hydriodic Acid, and Iodides	32
12. Detection of Bromides in the presence of Iodides	32
13. Detection of Chlorides in the presence of Iodides	32
14. Separation of an Iodide from a Bromide and Chloride	33
15. Iodates	33
15a. Detection of an Iodate in an Iodide	33
16. Periodates	33
17. Water and Hydrates	33
18. Oxides	34
19. Sulphur, Hydrosulphuric Acid, and Sulphides	34
20. Detection of a Soluble Sulphide in presence of a Sulphite and a Sulphate	36
21. Thiosulphates (Hyposulphates)	36
22. Separation of Thiosulphates from Sulphides	36
23. Sulphurous Acid and Sulphites	36
24. Sulphuric Acid and Sulphates	37
25. Separation of Sulphides, Sulphites, and Sulphates	38
26. Carbon, Carbonic Acid, and Carbonates	38
27. Boric Acid and Borates	39
28. Silicic Acid and Silicates	39
29. Separation of Silicic Anhydride (Silica) from all other Acids	40
30. Hydrofluosilicic Acid	40
31. Nitrous Acid and Nitrites	40
32. Detection of a Nitrite in the presence of a Nitrate	41
33. Nitric Acid and Nitrates	41
34. Detection of free Nitric Acid in the presence of a Nitrate	42
35. Detection of a Nitrate in the presence of an Iodide	42
36. Separation of Chlorides, Iodides, and Bromides from Nitrates	42
37. Cyanogen, Hydrocyanic Acid, and Cyanides	43
38. Separation of Cyanides from Chlorides	44
39. Cyanic Acid and Cyanates, Cyanuric Acid, and Fulminic Acid	44
40. Thiocyanates (Sulphocyanates)	44
41. Ferrocyanides	44
42. Ferricyanides	45
43. Separation of Ferro- from Ferricyanides	45
44. Detection of Cyanides in the presence of Ferro- and Ferricyanides	45
45. Hypophosphites	46
46. Phosphorous Acid and Phosphites	46
47. Meta- and Pyro-Phosphoric Acids and their Salts	46
48. Orthophosphoric Acid (B.P.) and Orthophosphates	47
49. Detection of a Phosphate in the presence of Calcium, Barium, Strontium, Manganese, and Magnesium	48
50. Detection of a Phosphate in the presence of Iron	48
51. Arsenious Acid and Arsenites	48
52. Arsenic Acid and Arseniates	48
53. Separation of an Arseniate from a Phosphate	49
54. Manganates	49
55. Permanganates	49
56. Chromic Acid and Chromates	49
57. Stannic Acid and Stannates (Stannites ?)	50
58. Antimonic Acid	50
59. Formic Acid and Formates	50
60. Detection of a Formate in presence of Fixed Organic Acids which reduce Silver Salts	50
61. Acetic Acid and Acetates	51
62. Valerianic Acid and Valerianates	51
63. Sulphovinates (Ethysulphates)	51
64. Stearic Acid and Stearates	52
65. Oleic Acid and Oleates	52
66. Lactic Acid and Lactates	52
67. Oxalic Acid and Oxalates	52
68. Succinic Acid and Succinates	53
69. Malic Acid and Malates	53
70. Tartaric Acid and Tartrates	53
71. Citric Acid and Citrates	54
72. Separation of Oxalates, Tartrates, Citrates, and Malates	55
73. Meconic Acid and Meconates	55
74. Carbolic Acid (or Phenol) and Carbolates (Phenates)	55
75. Benzoic Acid and Benzoates	56
76. Salicylic Acid	56
77. Detection of Carbolic Acid in presence of Salicylic Acid	57
78. Tannic, Gallic, and Pyrogallic Acid	57

CHAPTER IV.

Qualitative Analysis, as applied to the Detection of Unknown Salts.

	PAGE
§ I. General Preliminary Examination	58
§ II. Detection of the Metal present in any Simple Salt (with tables for same)	61–64
§ III. Detection of the Metals in Complex Mixtures of two or more Salts (with Tables for same)	65
§ IV. Detection of the Acidulous Radical	75–80
Div. A. Preliminary Examination	75
" *B.* Preparation of Solution	77
" *C.* Course for Inorganic Acids	78
" *D.* Course for Organic Acids	80
Solubility Tables	82, 83
§ V. Special Processes for proving the Identity of certain readily recognisable Substances	84–88

TABLES.

	PAGE
Full table for the Detection of the Metal in a Solution containing one Base only	62
Table for the Detection of the Metal in a Simple Salt, the Metal present being limited to the Salts used and included in the British Pharmacopœia	64
Table for the Separation of Metals into Groups	66
Table *A.* Separation of Metals of Group I.	67
Table *B.* Separation of Metals of Group II., Div. (*a*)	68
Table *C.* Separation of Metals of Group II., Div. (*b*)	69
Table *D.* Separation of Metals of Group III., Div. (*a*)	70
Table *E.* Separation of Metals of Group III., Div. (*a*). in presence of Phosphoric Acid	71
Table *F.* Separation of Metals of Group III., Div. (*b*)	72
Table *G.* Separation of Metals of Group IV.	73
Table *H.* Separation of Metals of Group V.	74

CHAPTER V.

Qualitative Analysis of Alkaloids and of the so-called "Scale" Medicinal Preparations containing them, with a General Sketch of Toxicological Analysis.

		PAGE
Division A.	Course for the Detection of the Alkaloids and Alkaloid Salts used in Medicine	89–91
" *B.*	Qualitative Analysis of Scale Preparations	91–92
" *C.*	General Sketch of the Method of Testing for Poisons in Mixtures	92–93
" *D.*	General *Resumé* of the Tests for all the chief Alkaloids (as under)	94

Aconitine.	Colchicine.	Narceïne.	Sabadilline.
Atropine.	Colchiceïne.	Narcotine.	Sabatrine.
Beberine.	Coniine.	Nepaline.	Solanine.
Berberine.	Curarine.	Nicotine.	Staphysagrine.
Brucine.	Delphinine.	Papaverine.	Strychnine.
Caffeine.	Delphinoidine.	Physostigmine.	Taxine.
Calabarine.	Emetine.	Pilocarpine.	Thalictrine.
Chelidonine.	Gelsemine.	Piperine.	Theobromine.
Cinchonine.	Hyoscyamine.	Quinamine.	Thebaine.
Cinchonidine.	Jervine.	Quinidine.	Veratrine.
Codeine.	Morphine.	Quinine.	Veratroidine.

PART II.—QUANTITATIVE ANALYSIS.

CHAPTER VI.

Weighing, Measuring, and Specific Gravity.

	PAGE
1. Weighing and Measuring	95–96
2. Specific Gravity	
(a) Of Fluids	97–99
(b) Of Solids	99–100
(c) Practical Applications of Specific Gravity	101
(d) Of Gases	101
(e) Vapor Density	102–104

CHAPTER VII.

Volumetric Quantitative Analysis and Use of the "Nitrometer."

I. Introductory Remarks	105
A. Standard solutions	105
B. Indicators	105
C. General *Modus Operandi*	105
D. Apparatus employed	106
E. Weighing operations	107
II. Standard Solution of Oxalic Acid	108
A. Preparation and Check	108
B. Estimation of Alkaline Hydrates	108
C. Estimation of Alkaline Carbonates	109
D. Estimation of Lead	110
E. Estimation of the Organic Salts of the Alkalies	110
F. Official Standards of Strength	111
III. Standard Solution of Sodium Hydrate	112
A. Preparation and Check	112
B. Acidimetry	112
C. Official Standards of Strength	112
IV. Standard Solution of Argentic Nitrate	113
A. Preparation and Check	113
B. Estimation of Soluble Haloid Salts	113
C. Volhard's Method	113
D. Estimation of Hydrocyanic Acid	114
E. Official Standards of Strength	114
V. Standard Solution of Iodine	114
A. Preparation and Check	114
B. Estimation of Arsenious Acid	115
C. Estimation of Sulphurous Acid	115
D. Estimation of Thiosulphates	115
VI. Standard Solution of Sodium Thiosulphate	115
A. Preparation and Check	115
B. Estimation of Free Iodine	115
C. Estimation of Free Chlorine and Bromine	116
D. Estimation of Available Chlorine	116
E. Official Standards of Strength	117
VII. Standard Solution of Potassium Bichromate	117
A. Preparation and Check	117
B. Estimation of Ferrous Salts	117
C. Estimation of Ferric Salts	118
D. Official Standards of Strength	118
VIII. Fehling's Standard Solution of Copper	119
A. Manufacture and Check	119
B. Estimation of Sugar	120
C. Estimation of Starch	120
IX. Estimation of Phosphoric Acid	120
X. Estimation of Sulphates by Standard Barium Chloride	121
XI. Standard Mayer's Solution for Estimation of Alkaloids	121
XII. Analysis by the Nitrometer	122
A. General Remarks	122
B. Estimation of Nitrous Ether	123
C. Estimation of Nitrates	123
D. Estimation of Soluble Carbonates	123
E. Estimation of Hydrogen Peroxide	124
F. Estimation of Urea	124
XIII. Colorimetric Analysis	124
(a) Estimation of Ammonia by "Nesslerising"	125
(b) Estimation of Salicylic Acid	126
XIV. Table of Coefficients for Volumetric Analysis	127

CHAPTER VIII.

Gravimetric Quantitative Analysis of Metals and Acids.

	PAGE
Div. I. Preliminary Remarks	128
A. Preparation of Filters	128
B. Estimation of Ash Filters	129
C. Collection and Washing of Precipitates	129
D. Drying of Precipitates	129
E. Igniting and Weighing of Precipitates	130
F. Estimation of Moisture	131
G. „ of Ash in Organic Bodies	131
H. Use of Analytical Factors	131
Div. II. Gravimetric Estimation of the Metals	132
1. Estimation of Silver:	
A. As Chloride	132
B. As Metal	132
2. Estimation of Lead:	
A. As Oxide	132
B. As Chromate	133
3. Estimation of Mercury:	
A. As Metal	133
B. As Sulphide	133
4. Estimation of Cadmium	134
5. „ of Copper:	
A. As Oxide	134
B. As Metal	134
6. Estimation of Bismuth:	
A. As Oxide	134
B. As Sulphide	135
7. Estimation of Gold	135
8. „ of Platinum	135
9. „ of Tin:	
A. As Oxide	135
B. As Metal	135
10. Estimation of Antimony	136
11. „ of Arsenic:	
A. As Sulphide	136
B. As Magnesium-Ammonium-Arseniate	136
12. Estimation of Cobalt	136
13. „ „ Nickel	137
14. „ „ Manganese	137
15. „ „ Zinc	137
16. „ „ Iron	137
17. „ „ Aluminium	138
18. „ „ Chromium	138
19. „ „ Barium	138
20. „ „ Calcium	138
21. Estimation of Magnesium	139
22. „ „ Potassium	139
23. „ „ Sodium	139
24. „ „ Potassium and Sodium in presence of Metals of the Fourth Group	140
25. Estimation of Ammonium	140
Div. III. Gravimetric Estimation of Acidulous Radicals	141
1. Chlorides	141
2. Iodides	141
3. Bromides	141
4. Cyanides	141
5. Estimation of an Iodide in the presence of a Chloride and Bromide	141
6. Mutual Estimation of Chloride, Bromide, and Iodide in the presence of each other	141
7. Sulphides	141
8. Sulphates	142
9. Nitrates	
A. In Alkaline Nitrates	142
B. As Nitric Oxide	142
C. As Ammonia	143
10. Estimation of Phosphates	
A. Estimation of the strength of Free Phosphoric Acid	143
B. Alkaline Phosphates	143
C. In presence of Calcium and Magnesium	143
D. In presence of Iron and Aluminium	144
E. Estimation as Phospho-molybdate	144
11. Estimation of the "Total" and "Soluble" Phosphates in a Manure or Soil	144
12. Estimation of Arseniates	145
13. „ „ Carbonates	145
14. „ „ Oxalic Acid	146
15. „ „ Tartaric Acid	146
16. „ „ Silicic Acid	146
A. Insoluble Silicates	146
B. Soluble Silicates	146
Div. IV. Quantitative Separations: Full Mineral Analysis of Water	147

CHAPTER IX.

Ultimate Organic Analysis.

1. Apparatus required	150
2. Estimation of Carbon and Hydrogen	151
3. Estimation of Nitrogen	153
(*a*) Method of Varrentrapp and Will	153
(*b*) Process of Dumas	154
(*c*) Kjeldahl's Process	154
4. Estimation of Chlorine	154
5. Estimation of Sulphur and Phosphorus and Arsenic	155

CHAPTER X.

Special Processes for the Analysis of Water, Air, and Food.

	PAGE
Div. I. The Sanitary Analysis of Water	156
1. Collection of Sample	156
2. Color	156
3. Odor	156
4. Total Solids	156
5. Chlorine	157
6. Phosphoric Acid	157
7. Nitrogen and Nitrates	157
8. Ammonia and Albuminoid Ammonia	158
9. Oxygen Consumed	160
10. Hardness	161
11. Judging the Results	162
Div. II. The Sanitary Analysis of Air	163
1. Estimation of Carbon Dioxide	163
2. Estimation of Organic Matter	163
3. Testing for Gaseous Impurities	163
Div. III. Food Analysis	164
1. Milk	164
Table Degrees of Thermometer	165
2. Butter	166
3. Bread	167
4. Alcohol (Estimation of)	167
Table Percentages of Alcohol	168
5. Mustard	169
6. Pepper	169
7. Colored Sweets	169
8. Direct Estimation of Starch (new process)	169
9. Vinegar	171

CHAPTER XI.

Special Processes for the Analysis of Drugs, Urine, and Urinary Calculi.

	PAGE
Div. I. Analysis of Drugs	172
1. General Scheme of Dragendorff	172
2. Cinchona Bark	174
3. Opium	176
4. Alkaloidal Strength of Extracts	177
5. Methylated Spirit in Tinctures	177
6. Strength of Resinous Drugs	178
7. Testing the Purity of Quinine Sulphate	178
8. Alkaloidal Strength of Scale Preparations	179
9. Estimation of Phenol	179
10. Estimation of Fatty Acids in Soap	180
11. Estimation of Oleic Acid	180
12. „ of Glycerine	180
Div. II. Micro-chemical Analysis of Drugs (*illustrated*)	181
Div. III. Analysis of Urine	184
1. Specific Gravity	184
2. Reaction	184
3. Deposit	184
4. Albumin (and estimation)	184
5. Sugar (and estimation)	185
6. Bile	186
7. Urea (and estimation)	186
8. Uric Acid „	186
9. Phosphates	187
10. Sulphates	187
11. Chlorides	187
12. Blood	187
Analysis of Urinary Calculi:	
(*a*) Organic	188
(*b*) Inorganic	188

CHAPTER XII.

Analysis of Gases, Polarisation, and Spectrum Analysis.

Div. I. Analysis of Gases by Keiser's Apparatus	189
„ II. Analysis by the Polariscope	192
„ III. Spectrum Analysis	194
„ IV. Melting Points	195

PART I.

QUALITATIVE ANALYSIS.

CHAPTER I.

THE PROCESSES EMPLOYED BY PRACTICAL CHEMISTS.

It is advisable that the student should understand the *raison d'être* of the chief processes he will be called upon to employ, before commencing in detail the study of Analysis.

I. SOLUTION.

This process consists in stirring a solid body in contact with a fluid until it dissolves, heat being occasionally used. Bodies which refuse to dissolve in any particular fluid are said to be **insoluble** in it; the liquid used is called the **solvent**, and sometimes the **menstruum**; a liquid having taken up all the solid matter possible is said to be **saturated**. A knowledge of the solubility of various substances in the chief menstrua, such as water, acids, alkalies, alcohol, ether, chloroform, and glycerine, is of the utmost importance. By this we are enabled to separate one body from another; and by attention to minute details it is even possible to divorce bodies which are soluble in the same menstruum, but in different degrees. As an example we shall suppose three substances—one readily soluble, one partially so, and one but slightly soluble, in ether. By a careful use of three successive quantities of the liquid at different temperatures, we can obtain separate solutions of each. This process is named *fractional solution*. In order to ascertain if any substance be soluble in any particular liquid, it is simply requisite to stir it into the fluid, applying heat if necessary. A portion of the liquid is then poured off, and heated until it has passed away in vapour, when, if any of the solid was held in solution, it will remain as a visible residue. As a general rule, the higher the temperature to which a liquid is raised, the greater becomes its capacity for saturation. Thus, one part of common nitre will require for solution about four parts of cold water; but the same weight of the solid will dissolve in less than half a part of boiling water. There are many exceptions to this rule—notably that of calcium oxide, which is less soluble in boiling water than in cold. Many bodies, during solution, absorb so much heat that any substance placed in the liquid has its temperature remarkably reduced. Potassium nitrate and ammonium chloride, dissolved together in water, form in this manner a very efficient refrigerant when snow or ice is not obtainable.

II. LIXIVIATION AND EXTRACTION.

These processes include the digestion of a mixture of solids in a fluid, so as to dissolve the soluble portion. The solids, in the form of powder, are introduced into a vessel, and the water or other liquid having been added, the whole is well stirred. Remaining at rest until all the insoluble matter has subsided, the clear fluid, charged with all that was soluble in the mixture, is decanted, or poured off. On the large scale, in chemical works, the clear liquor is usually drawn off by means of a **siphon.**

When a substance is to be extracted by means of a readily volatile solvent, such as ether or chloroform, the arrangement now commonly used is that known as **Soxhlet's apparatus.** This is illustrated in fig. 1. A flask (A) is charged with the solvent. The substance (E) is put into a cartridge of filtering paper, and introduced into the Soxhlet tube (D); the latter is in turn connected with the upright condenser (C), through the jacket of which a stream of cold water is made to pass. Heat is now applied to the flask by a water-bath, and the vapor of the ether rising through (B), condenses and drops on to the powder in the cartridge. When the instrument has become filled by the condensed solvent to the level of the top of (F), it runs back into the flask charged with the soluble matter that has been extracted. This process then repeats itself until the whole soluble portion has been extracted and transferred to the flask, which latter may then be attached to an ordinary condenser, and the solvent distilled off, leaving the soluble matters of the original powder in the flask. Resinous and sticky substances should be mixed with a little purified sand to prevent them clogging up the apparatus.

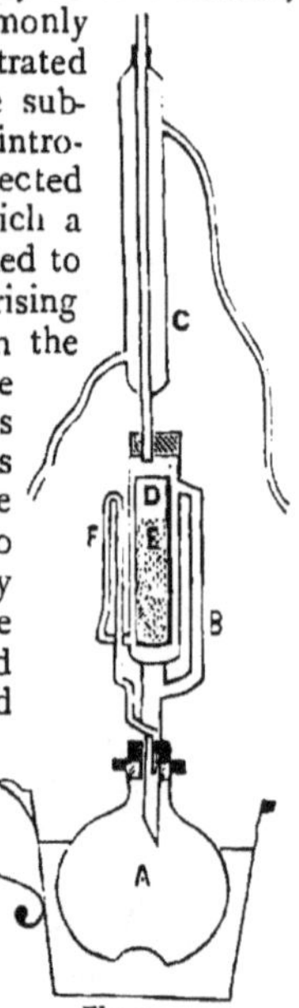

Fig. 1.

Another method of extraction is that known as **percolation,** a process much used in pharmacy. The apparatus employed is illustrated in fig. 2. The upper portion (A) is the percolator, in which the powder to be extracted is tightly packed; and the solvent having been poured upon it, the whole is allowed to macerate for some time. The stopcock (C) being then opened, the fluid gradually filters into the receiver (B), and more solvent is then poured on until the soluble portion of the contents of the percolator has been entirely transferred to the receiver.

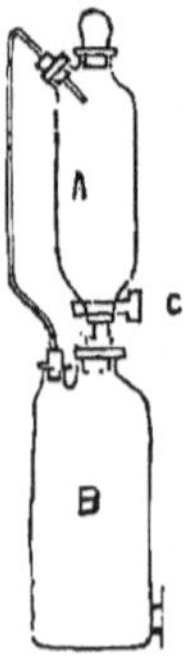

Fig. 2.

III. PRECIPITATION

Is the mixing of two substances in solution so as to form a third substance, which, being insoluble in the fluids employed, sinks to the bottom, and is called the **precipitate.** The clear liquid which remains after the precipitate has settled down, is called the **supernatant liquid.**

When precipitates are totally insoluble in water, such as barium sulphate or argentic chloride, the operation is best conducted at a boiling heat, as the high temperature causes the precipitate to aggregate and become more dense, so that it subsides rapidly, and is less liable to pass through the pores of the filter.

On the other hand, there are some precipitates which must never be heated, but be allowed to form slowly by standing in the cold for several hours. To this class belong ammonium-magnesium phosphate, and potassium acid tartrate. When it is desirable to cause precipitates to form quickly, resort may be had to violent shaking, or stirring with a glass rod, so that it scrapes against the sides of the vessel.

In qualitative analysis precipitation is usually conducted in **test-tubes**, shown in fig. 3. These are kept in a stand (fig. 4) having holes for the reception of tubes actually in use, and also a row of pegs upon which freshly

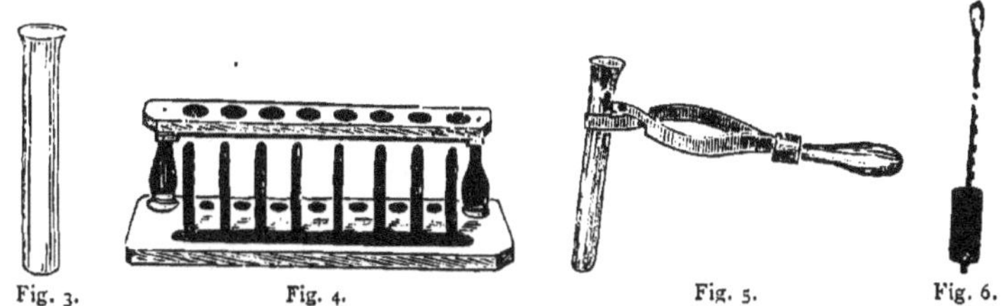

Fig. 3. Fig. 4. Fig. 5. Fig. 6.

washed tubes may be inverted to drain. Figure 5 illustrates the appliance used to hold tubes when their contents are to be boiled, and fig. 6 shows the brush used for cleansing them.

In quantitative analysis precipitation is generally performed in **beakers**. These are very thin tumblers made of glass, free from lead, and annealed so as to permit their use with boiling liquids without risk of fracture. A nest of such beakers is illustrated in fig. 7.

Fig. 7.

Precipitates are separated from the supernatant liquor by one of two methods. First :—

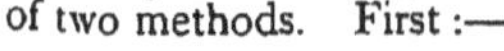

IV. DECANTATION,

which consists in allowing the precipitate to settle to the bottom of the liquid; pouring off the clear liquor, then pouring on distilled water, and repeating the process until the precipitate is thoroughly washed. Second :—

V. FILTRATION,

which consists in transferring the whole to a piece of folded filtering-paper or cloth placed in a funnel, so that the liquid passes through, while the precipitate remains. This may be washed by the addition of distilled water. The liquor which has thus passed through the filter is called the **filtrate**. The process of precipitation is probably one of the most common with which the practical chemist meets.

Specially prepared papers for filtration, cut into circles, are sold, and they have only to be folded to fit the funnel for which they are destined. The

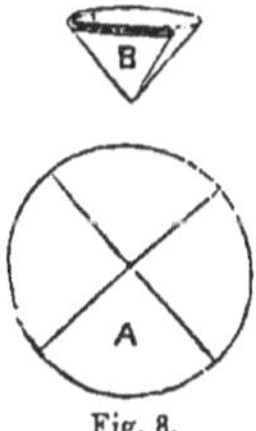

Fig. 8.

Fig. 9.

illustrations show, fig. 8 (A), the circle of paper, and (B) the same as folded for use, while fig. 9 represents the whole arrangement ready for use.

VI. DISTILLATION

is the changing of a fluid into vapour by the aid of heat, and passing the vapour into a cooling apparatus, called the **condenser**, where, its latent heat being abstracted, it is again deposited as a liquid. This process is employed for the separation of volatile fluids from non-volatile substances. The fluid which passes over and is condensed in the receiver is called the **distillate**; while the non-volatile matter which remains in the retort is called the **residue**. Figure 10 shows the arrangement most commonly used in laboratories for small distillations. A is the **retort** in which the fluid is boiled, and B is called a **Liebig's condenser**, through the jacket of which a stream of water is caused to pass, and beneath the end of this is placed a vessel (C) for the reception of the distillate. By careful attention to their boiling points, various volatile fluids may thus be separated from each other. Suppose, for example, that we have a mixture of three substances boiling respectively at 180°, 220°, and 240°, and their separation is desired, we should introduce the mixture into a retort fitted with a thermometer, the bulb of which was placed just above the level of the fluid. The whole being then attached to the condenser, the heat would be gradually raised until the thermometer marked a little over 180°, when that temperature would be steadily maintained as long as anything continued to collect in the receiver. When the liquid ceased to accumulate, the receiver would be changed, and the temperature raised to a little over 220°, and the distillation continued until the second liquid had ceased to pass over. The receiver being once more changed, the heat would be again raised, and maintained until the last liquid had been obtained as a distillate. This process, which is exceedingly useful in practical chemistry, is called **fractional distillation.**

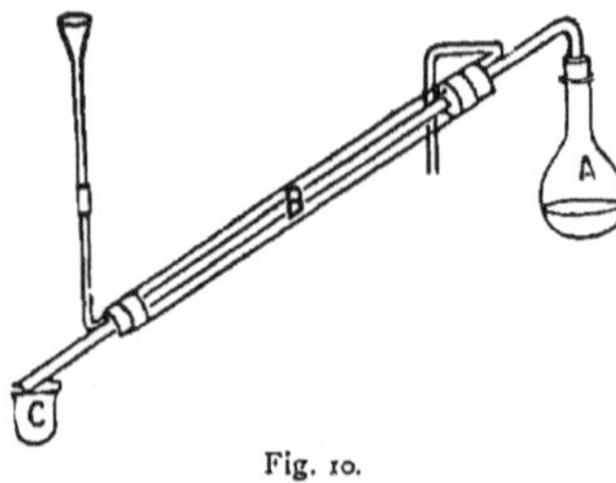

Fig. 10.

VII. SUBLIMATION

is the changing of a solid into a vapor by heat, and recondensing the vapor into a solid form in a cooled vessel. It is employed for the separation of volatile from non-volatile solids, and is thus conducted :—The substance to be sublimed is thinly spread over the bottom of a shallow iron pan, and the vessel is covered with a sheet of bibulous paper perforated with numerous pin-holes, or with a piece of muslin. By means of a sand bath, the heat is slowly raised to the desired degree, when the vapor passes through the strainer, and condenses in a cap of wood or porcelain, lined with stout cartridge paper, and placed over the heating-pan and kept cool. **Fractional sublimation** is often useful, and may be employed in a similar manner to fractional distillation.

VIII. FUSION

is the heating of a solid until it melts. It is usually carried out in a vessel called a "crucible," made of fire-clay. On the small scale, or for the purposes of analysis, fusion is generally conducted in porcelain crucibles; or, where the substances are such as would attack porcelain, in vessels of platinum or silver. Alkalies should be fused only in crucibles made of the latter metal. A peculiar kind of fusion, called **cupellation**, is resorted to in the assay of gold and silver. The alloy to be assayed is wrapped in a piece of lead foil, and the whole is then heated in a little cup made of bone earth, called a **cupel**, when the lead and all impurities fuse and sink into the substance of the porous vessel, leaving the pure metal as a metallic button, which may then be weighed. The illustration (fig. 11) shows a set of crucibles for fusion—A, being of fire-clay; B, a platinum crucible; C, one of porcelain; and D, what is called *Rose's crucible*, for heating substances in a current of hydrogen when

Fig. 11.

it is desired to prevent access of air, or to produce rapid reduction to the metallic state.

IX. EVAPORATION

consists in heating a fluid until the whole, or as much of it as may be required, passes off in vapor. A solution thus treated until it has wholly passed into vapor is said to be **evaporated to dryness**, and the solid substance remaining is called the **residue**.

Solutions which contain any organic or volatile matter ought always to be evaporated on a water bath; that is, in a vessel exposed only to the heat of boiling water. With ordinary non-volatile or metallic substances in solution this precaution is unnecessary, and of no practical advantage. Evaporation may be conducted slowly, without raising the fluid to its boiling point, when it is called simple **vaporisation**; but when sufficient heat is applied, the evaporation takes place rapidly, and is accompanied by the disengagement of bubbles of vapour, and the fluid is then said to be in a state of **ebullition.** All liquids possess the continual desire, as it were, to pass into vapor; and the amount of elastic force which the vapor thus given off exerts is called its **tension.** The more the liquid is heated, the greater becomes its tendency to vaporise, and consequently the more powerful is the tension of its vapor; and when the latter is sufficiently marked to overcome the pressure exerted by the atmosphere and the cohesion of the liquid itself, ebullition takes place. The boiling point of a fluid is therefore *the temperature at which the tension of its vapor is equal to that of the superincumbent atmosphere.* If the pressure of the atmosphere be increased artificially, the boiling point of the liquid will rise in proportion. For example: boiling water in an open vessel under ordinary circumstances will have a temperature of 212°; but water in a steam boiler, under a pressure of 60 lb. per square inch, will be found to be heated nearly to 264°. As steam under pressure can thus be obtained at high temperatures, it is made use of for the rapid evaporation of liquids on a large scale, by causing it to pass into a jacket or through a coil of pipes surrounding the evaporating pan. The apparatus thus made use of is called a *steam bath*, the heat of which is usually understood to be about 230° Fahr.

If water be boiled in a chemically clean glass vessel, and more particularly if a precipitate be suspended in the water, the boiling does not take place regularly, but the liquid becomes heated above its boiling point, and suddenly rushes into vapor in gusts. This is called by practical chemists "bumping," and may be prevented by putting in a few fragments of platinum foil, which act as *nuclei*, to aid in the regular disengagement of the vapor.

X. CRYSTALLISATION AND DIALYSIS.

Many substances when dissolved in a boiling liquid separate out, as soon as the fluid cools, in masses having a well-defined and symmetrical shape, bounded by plain surfaces and regular angles. These bodies are named **crystalline**; the deposited masses, **crystals**; and the remaining solution, the **mother liquor**. Substances which are not susceptible of crystallisation are called **amorphous** (*i.e.* formless) bodies; while solids, such as glue and gums, which are soluble in water and yet not crystallisable, are named **colloids**. Crystallisation may also occur during solidification after fusion, and by the spontaneous evaporation of liquids holding crystalline substances in solution. All crystalline bodies invariably assume the same form, and may thus be unmistakably recognised from each other. The process is also useful for purification, as at the moment of crystallisation all impurities are rejected, and may be poured off with the mother liquor. Many circumstances affect the size of the crystals produced in any solution; as a rule, the more rapidly crystallisation takes place, the smaller are the crystals. An example of the extreme variation in the size of the crystals produced from the same solution may be seen in ferrous sulphate. When allowed to deposit slowly, we have the ordinary well-marked commercial crystals; but the same salt dissolved in boiling water, and the solution suddenly poured, with constant stirring, into spirit, gives granulated ferrous sulphate in crystals so minute that a lens is required to distinguish them. For the formation of large and well-defined crystals perfect rest is required, and it is often desirable to introduce pieces of wood or string so as to form nuclei on which the crystals collect. Good examples are seen in commercial crystallised sugar-of-milk and sugar-candy. Some bodies are capable of crystallising in two or more forms, and are called **polymorphous**. Instances of this property may be seen in mercuric iodide and in sulphur.

When small quantities of crystalline substances exist in a solution together with a large quantity of uncrystallisable colloid bodies, their mutual separation is effected by **dialysis**. This process consists in introducing the mixture into a glass vessel having a bottom made of vegetable parchment. This, called the **dialyser**, is floated in a large quantity of distilled water in a basin. At the expiration of several hours the whole of the crystalline bodies will have passed through the parchment, and will have become dissolved in the water in the basin, while the colloids remain in the dialyser. A very good way of practising this process is to dialyse a solution of glue in which a few grains of salt have been dissolved, when the salt will be found to have passed into the water in the basin, while all the glue will remain behind. This process is sometimes employed for the separation of crystalline poison, like strychnine, from the contents of a stomach. The rapidity of the dialysis is greatly increased by causing a stream of water to pass through the outer basin, but of course this is only applicable where it is desired to retain the colloid body and not the crystalloid, as in the manufacture of the preparation known as *dialysed iron.* The apparatus used is shown in the illustration (fig. 12), in which A is the dialyser containing the mixture, while B contains the water into which the crystalline matter passes.

Fig. 12.

XI. ELECTROLYSIS.

Electricity is one of the most powerful analytic forces with which the chemist has to deal. A current of electricity results from the action of chemical agents upon various metals, the apparatus in its simplest form being called a **voltaic** or **galvanic cell** or **element**, and a combination of several cells being termed a **battery**. Bunsen's battery, which is very extensively used and of great power, consists of an outer cell or jar of glazed earthenware in which is placed a cylinder of zinc, an inner unglazed porous jar and a rod of carbon, both furnished with binding-screws for attaching wires to them. The acids used are dilute sulphuric around the zinc, and strong nitric in contact with the carbon in the inner cell. The zinc constitutes the positive (or most attacked), the carbon the negative (or least attacked) element; but when wires are attached to these, the end of the wire connected with the zinc is termed the **negative electrode** or **cathode**, whilst that in union with the platinum in the battery is the **positive electrode** or **anode**. To explain this apparent contradiction we must imagine that the current generated by the zinc passes through the fluid in the cell and then through the carbon, while at the same time an opposite current is generated from the carbon through the liquid to the zinc, and thus the effect of each current is felt at the ends of the wires from the opposite plate. These electrodes, when placed near one another in a compound liquid, occasion the phenomena of **electrolysis**, which is simply the splitting up of conductors of electricity into their elements or into simpler forms. For instance, a solution of HCl gives off Cl at the positive electrode and the H at the negative, as a result of the axiom that unlike electricities attract and like repel one another, Cl itself being electro-negative and H electro-positive. Dealing with H_2SO_4, we get H_2 at one electrode and SO_4 at the other; the latter, however, at once splitting up into O (given off) and SO_3, which re forms H_2SO_4 with the water present.

Further reference to applications of electrolysis will be found in Chapter X., when the student arrives at quantitative analysis. It is also applied in a modified form in qualitative analysis to the separation of tin and antimony.

XII. PYROLOGY.

Under this name are included all processes of analysis depending for their action on the use of fire, or, in other words, what were formerly called "*reactions in the dry way*." The chief instruments used are the blowpipe and the Bunsen burner. The **blowpipe** is a tube with a narrow nozzle, by which a continuous current of air can be passed into an ordinary flame. The ordinary gas flame consists of three parts: (*a*) A non-luminous nucleus in the centre; (*b*) A luminous cone surrounding this nucleus; and (*c*) An outer and only slightly luminous cone surrounding the whole flame. The centre portion (*a*) contains unaltered gas, which cannot burn for want of oxygen, that necessary element being cut off by the outer zones. In the middle portion (*b*) the gas comes in contact with a certain amount of oxygen, but not enough to produce complete combustion; and therefore it is chiefly the hydrogen which burns here, the carbon separating and, by becoming intensely ignited, giving the light. In the outer zone (*c*) full combustion takes place, and the extreme of heat is arrived at, because chemical action is most intense. The outer flame therefore acts readily on oxidisable bodies, because of the high temperature and the unlimited supply of air, while the luminous zone tends to take away oxygen by reason of the excess of unburned carbon or hydrocarbons therein existing. For these reasons the former is called the **oxidising flame**, and the latter the **reducing flame**. The effect of blowing air across a flame is, first, to alter the shape of the flame, which is at once lengthened and narrowed; and, in the second place, to extend the sphere of combustion from the outer to the inner part. As the latter circumstance causes an increase of the heat of the flame, and the former a concentration of that heat within narrower limits, it is easy to understand the great heat of the blowpipe flame. The way of holding the blowpipe and the strength of the blast always depends upon whether the operator wants a *reducing* or an *oxidising* flame. The *reducing* flame is produced by keeping the jet of the blowpipe just on the border of a tolerably strong gas flame, and driving a moderate blast across it. The resulting mixture of the air with the gas is only imperfect, and there remains, between the inner bluish part of the flame and the outer barely visible part a luminous and reducing zone, of which the hottest point lies somewhat beyond the apex of the inner cone. To produce the *oxidising* flame, the gas is lowered, the jet of the blowpipe pushed a little farther into the flame, and the strength of the current somewhat increased. This serves to effect an intimate mixture of the air and gas, and an inner pointed, bluish cone, slightly luminous towards the apex, is formed, and surrounded by a thin, pointed, light-bluish, barely visible mantle. The hottest part of the flame is at the apex of the inner cone. Difficultly fusible bodies are exposed to this part to effect their fusion; but bodies to be oxidised are held a little beyond the apex, that there may be no want of air for their combustion.

The *current* is produced by the cheek muscles alone, and not with the lungs. The way of doing this may be easily acquired by practising for some time to breathe quietly with puffed-up cheeks and with the blowpipe between the lips; with practice and patience the student will soon be able to produce an even and uninterrupted current.

The *supports* on which substances are exposed to the blowpipe flame are generally either wood charcoal, or platinum wire or foil.

Charcoal supports are used principally in the reduction of metallic oxides, etc., or in trying the fusibility of bodies. The substances to be operated upon are put into small conical cavities scooped out with a penknife or with a little tin tube. Metals that are volatile at the heat of the reducing flame evaporate wholly or in part upon the reduction of their oxides; in passing through the outer flame the metallic fumes are re-oxidised, and the oxide formed is deposited around the portion of matter upon the support. Such deposits are called incrustations. Many of these exhibit characteristic colors leading to the detection of the metals. Thoroughly burnt and smooth pieces of charcoal only should be selected for supports in blowpipe experiments, as imperfectly burnt and knotty pieces are apt to spirt and throw off the matter placed on them.

The great use of charcoal lies (1) in its low degree of conductivity; (2) its porosity, which causes it to absorb fusible bodies and leave infusible ones upon its surface; and (3) its power of aiding the effects of the reducing flame.

Platinum wire and **foil** are used for supports in the oxidising flame, and the former is specially employed for trying the action of fluxes and the color communicable to the blowpipe or Bunsen flame. The platinum wire, when employed for making beads of borax or other fluxes, should be about 3 to 4 inches long, with the end twisted into a small loop. The loop is then heated, and dipped while hot in the powdered borax, when it takes up a quantity which is then heated till it fuses to a clear bead formed within the loop. When cold, this is moistened and dipped in the powder to be tested, and again exposed to the flame, and the effect noted. For trying the color impartable to the flame by certain metals, the wire is first cleaned by boiling in dilute nitric acid and then holding it in the flame until no color is obtained. The loop is then dipped in the solution to be tested, and held near the flame till the adhering drop has evaporated to dryness, and then heated in the mantle of the flame near the apex of the inner cone, and the effect observed.

The **Bunsen burner** consists of a tube having at its base a series of holes to admit air, and also a small gas delivery tube. By means of this contrivance the gas is mixed with air before it burns, and more perfect oxidation, and consequently much greater heat, is secured. Looking attentively at the flame of a Bunsen burner, we distinguish in it an inner part and two mantles surrounding it. The inner part corresponds to the dark nucleus of the common gas flame, and contains the mixture of gas and air issuing from the burner.

The mantle immediately surrounding the inner part contains still some unconsumed carbide of hydrogen; the outer mantle, which looks bluer and less luminous, consists of the last products of combustion. This hottest part—lying in the mantles surrounding the inner part of the flame, in a zone extending a few hundredths of an inch upwards and downwards from the transverse section of the flame across the apex of the inner part—has, according to Bunsen's calculation, a temperature of 4172° F. This is termed the *zone of fusion*. The outer margin of this zone of fusion acts as *oxidising flame*, the inner part of it as *reducing flame*. The spot where the reducing action is the most powerful and energetic lies immediately above the apex of the inner part of the flame. The Bunsen flame brings out the coloration which many substances impart to flames, and by which the qualitative analyst can detect many bodies, even though present in such minute quantities that all other means of analysis except the spectroscope fail to discover them. The subject of the coloration of flames will be discussed fully under each metal.

XIII. PREPARATION OF SULPHURETTED HYDROGEN.

This is done by acting upon ferrous sulphide with dilute sulphuric acid. The illustration (fig. 13) shows the apparatus. The ferrous sulphide, broken into lumps the size of a nut, is placed in the generating bottle (A), and dilute sulphuric acid is poured in by the funnel (C) in small quantities as required. (B) is a bottle containing distilled water, through which the gas

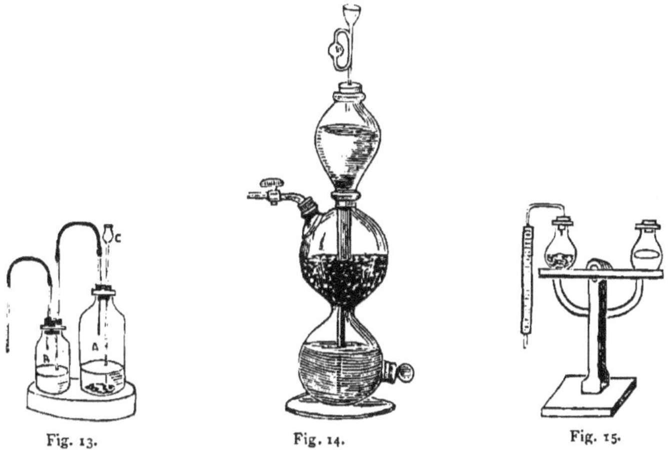

Fig. 13. Fig. 14. Fig. 15.

passes to free it from any traces of acid mechanically carried over. Owing to the disagreeable odor, it is desirable to have special appliances, by means of which the evolution of the gas can be stopped as soon as it has done the work required. Such an apparatus for use in a large laboratory is that of Kipps (fig. 14); and one suitable for use by a single student is that of Van Babo (fig. 15). Both illustrations sufficiently explain themselves.

CHAPTER II.

DETECTION OF THE METALS.

FOR the purposes of qualitative analysis we employ certain chemicals, either in the solid or liquid state, which by producing given effects enable us to detect the existence of the substance searched for. These substances are always kept ready for use, and are called *reagents*. They are of three classes: 1st, **Group reagents**, which, by yielding a precipitate under certain conditions, prove the substance to be a member of a certain group of bodies; 2nd, **Separatory reagents**, by means of which the substance under examination is distinguished from the other members of the group; 3rd, **Confirmatory reagents**, by which the indications previously obtained are confirmed and rendered certain.

The Metals are divided into five groups, each of which has its group reagent, as follows:

GROUP 1. Metals the chlorides of which, being insoluble in water, are precipitated from their solution by the addition of **hydrochloric acid**. They are **silver, mercurous mercury** and **lead** (the latter in cold strong solutions only).

GROUP 2. Metals the sulphides of which, being insoluble in dilute hydrochloric acid, are precipitated from their solutions by the addition of **sulphuretted hydrogen** in the presence of hydrochloric acid. This group includes **mercury, lead, bismuth, copper, cadmium, antimony, tin, gold, platinum,** and the metalloid **arsenic**, and is divided into two sub-groups, as follows:—

A. Metals the sulphides of which are insoluble in both dilute hydrochloric acid and ammonium sulphide. The precipitated sulphides separated by sulphuretted hydrogen are therefore **insoluble**, after washing, in **ammonium sulphide**. They are **mercury, lead, bismuth, copper** and **cadmium.**

B. Metals the sulphides of which, although insoluble in dilute acids, are dissolved by alkalies, and the precipitates from their solutions by **sulphuretted hydrogen** therefore dissolve in **ammonium sulphide**. They are **gold, platinum, tin, antimony** and **arsenic.**

GROUP 3. Embraces those metals the sulphides of which are soluble in dilute acids, but are insoluble in alkalies, and which consequently, having escaped precipitation in Group 2, are now in turn precipitated by **ammonium sulphide**. They are **iron, nickel, cobalt, manganese** and **zinc**. In this group are likewise included **aluminium, cerium** and **chromium**, which are precipitated as **hydrates** by the alkalinity of the **ammonium sulphide**. **Magnesium** would also be precipitated as hydrate, but, as that would be inconvenient at this stage, its precipitation is prevented by the addition of **ammonium chloride**, in which its hydrate is soluble.

GROUP 4. Comprises metals the chlorides and sulphides of which, being soluble, escape precipitation in the former groups, but the carbonates of which, being insoluble in water, are now precipitated by **ammonium carbonate.** They are **barium, strontium,** and **calcium. Magnesium** is not precipitated as carbonate, owing to the presence of the **ammonium chloride** already added with the sulphide in Group 3.

GROUP 5. Includes metals the chlorides, sulphides and carbonates of which, being soluble in **water** or in **ammonium chloride,** are not precipitated by any of the reagents already mentioned. They consist of **magnesium, lithium, potassium, sodium** and **ammonium.**

As the analytical grouping of the metals is undoubtedly one which is most important to the student for practical purposes, we shall adhere to this arrangement in giving the methods for their detection.

GROUP I.

Metals precipitable as **chlorides** by the addition of hydrochloric acid to their solutions.

I. SILVER (Ag).

(*a*) *WET REACTIONS.*

(To be practised upon a solution of argentic nitrate—$AgNO_3$.)

1. **HCl** (*1st group reagent*) or any soluble chloride gives a curdy white precipitate of argentic chloride—AgCl—insoluble in boiling nitric acid but instantly soluble in ammonium hydrate. It is also soluble in KCy, $Na_2S_2O_3$, and in strong solutions of soluble chlorides.
2. **KHO** or **NaHO** both produce a brownish precipitate of argentic oxide, Ag_2O, insoluble in excess. A similar effect is produced by the hydrates of barium, strontium and calcium.
3. $\mathbf{K_2CrO_4}$ gives a red precipitate of argentic chromate—Ag_2CrO_4—soluble in large excess of both nitric acid and ammonium hydrate; and therefore the solution should always be as neutral as possible.
4. $\mathbf{H_2S}$ and $\mathbf{NH_4HS}$, both produce black argentic sulphide—Ag_2S—insoluble in excess, both soluble in strong boiling nitric acid.
5. **KI** and **KBr** both produce curdy precipitates, the former yellow argentic iodide—AgI—insoluble in ammonium hydrate, and the latter argentic bromide—AgBr—dirty-white and slowly soluble in ammonium hydrate.
6. **KCy** gives a curdy-white precipitate of argentic cyanide—AgCy—readily soluble in excess, and also in boiling strong nitric acid.
7. Many organic salts, such as formates and tartrates, boiled with solutions of silver, precipitate the metal as a mirror on the tube.
8. Fragments of copper, zinc, iron and tin, introduced into a solution of silver, all precipitate the metal.

(*b*) *DRY REACTION.*

(To be practised on argentic oxide—Ag_2O.)

Mixed with sodium carbonate and heated on charcoal before the blowpipe, a bead of silver is formed, hard, glistening, and soluble in nitric acid, yielding solution of argentic nitrate to which the wet tests may be applied.

II. MERCUROSUM $(Hg_2)''$.

(*a*) *WET REACTIONS.*

(To be practised on a solution of mercurous nitrate—$(Hg_2)''(NO_3)_2$—prepared by acting upon a globule of mercury with cold and dilute nitric acid, so as not quite to entirely dissolve the metal.)

1. **HCl** (*1st group reagent*) gives a white precipitate of mercurous chloride,—$(Hg_2)''Cl_2'$—turned to black mercuros-ammonium chloride—$NH_2(Hg_2)''Cl'$—by ammonium hydrate. It is also insoluble in boiling water, but soluble in strong nitric acid, being converted into a mixture of mercuric chloride—$HgCl_2$—and mercuric nitrate—$Hg(NO_3)_2$.
2. **KHO** and **NaHO** both give black precipitates of mercurous oxide—Hg_2O—insoluble in excess.
3. **NH_4HO** produces a black precipitate of dimercuros-ammonium nitrate—$NH_2(Hg_2)''NO_3H_2O$—also insoluble in excess.
4. **$SnCl_2$** boiled with the solution causes a grey precipitate of finely divided mercury, which, if allowed to settle, and then boiled with hydrochloric acid and some more stannous chloride, aggregates into a globule.

(*b*) *DRY REACTION.*

(To be tried upon mercurous iodide.)

Mercurous compounds heated, break up into the corresponding mercuric salt, which sublimes, and metallic mercury, which also sublimes at a higher temperature.

III. LEAD (Pb).

(*a*) *WET REACTIONS.*

(To be practised on a solution of plumbic acetate—$Pb(C_2H_3O_2)_2$.)

1. **HCl** (*1st group reagent*) forms in cold strong solutions a white precipitate of plumbic chloride—$PbCl_2$—soluble in boiling water.
2. **H_2S** *after acidulation by* **HCl** (*2nd group reagent*) gives a black precipitate of plumbic sulphide—PbS—insoluble in ammonium sulphide. By treatment with boiling strong nitric acid it is decomposed, partly into plumbic nitrate, but chiefly into insoluble plumbic sulphate. It is entirely dissolved by hot dilute nitric acid with separation of sulphur.
3. **H_2SO_4** gives a white precipitate of plumbic sulphate—$PbSO_4$—slightly soluble in water, but rendered entirely insoluble by the addition of a little alcohol. It is decomposed by boiling strong hydrochloric acid, and is also freely soluble in solutions of ammonium acetate or tartrate, containing an excess of ammonium hydrate.
4. **K_2CrO_4** gives a yellow precipitate of plumbic chromate—$PbCrO_4$—insoluble in acetic and very dilute nitric acids, but soluble in strong boiling nitric acid.
5. **KI** gives a yellow precipitate of plumbic iodide—PbI_2—soluble in boiling water, and crystallising out on cooling in golden scales.
6. **KHO** and **NaHO** both cause white precipitates of plumbic hydrate—$Pb(HO)_2$—soluble in excess, forming potassium or sodium plumbates—K_2PbO_2 and Na_2PbO_2.
7. **NH_4HO** causes a white precipitate of a white basic nitrate—$Pb(NO_3HO)$—insoluble in excess.
8. **KCy** produces a white precipitate of plumbic cyanide—$PbCy_2$—insoluble in excess, but soluble in dilute nitric acid.

9. **Alkaline Carbonates** cause a precipitate of $(PbCO_3)_2Pb(HO)_2$—"white lead"—insoluble in excess, and also in potassium cyanide.
10. Fragments of zinc or iron in the presence of a little acetic acid cause the separation of metallic lead in crystalline laminæ.

(*b*) *DRY REACTION.*

(To be practised on red lead—Pb_3O_4, or litharge—PbO.)

Heated on charcoal in the inner blowpipe flame, a bead of metallic lead is formed, which is soft and malleable and soluble in dilute nitric acid. The solution thus obtained gives the wet tests for lead.

GROUP II.

Metals which are not affected by acidulation with hydrochloric acid, but are precipitated by passing sulphuretted hydrogen through the acidulated solution.

DIVISION A.

Metals which when precipitated by sulphuretted hydrogen as above, yield sulphide insoluble in ammonium sulphide.

I. MERCURICUM (Hg).

(*a*) *WET REACTIONS.*

(To be practised on a solution of mercuric chloride—$HgCl_2$.)

1. H_2S *after acidulation by* **HCl** (*2nd group reagent*) gives a black precipitate of mercuric sulphide—HgS—insoluble in ammonium sulphide and nitric acid, and only soluble in nitro-hydrochloric acid. Care must be taken that the sulphuretted hydrogen is passed really in excess, and that the whole is warmed gently, as unless this be done, the precipitate is not the true sulphide, but a yellowish-brown dimercuric sulpho-dichloride—Hg_2SCl_2. Although insoluble in any single acid, mercuric sulphide may be caused to dissolve in hydrochloric acid by the addition of a crystal of potassium chlorate.
2. **KHO** or **NaHO** both give a yellow precipitate of mercuric oxide—HgO—insoluble in excess.
3. NH_4HO produces a white precipitate of an insoluble mercur-ammonium chloride—$(NH_2Hg)Cl$—also insoluble in excess.
4. **KI** yields a red precipitate of mercuric iodide, soluble in excess both of the precipitant and the mercuric salt.
5. $SnCl_2$, boiled with a mercuric solution, first precipitates mercurous chloride, and then forms metallic mercury, as in the case of mercurosum compounds.
6. **Alkaline Carbonates** (except ammonium carbonate) produce an immediate reddish-brown precipitate of mercuric oxy-carbonate.
7. Fragments of Cu, Zn, or Fe precipitate metallic mercury in the presence of dilute hydrochloric acid.

(*b*) *DRY REACTION.*

(To be tried on mercuric oxide—HgO—and on "Ethiops mineral"—HgS.)

All compounds of mercury are volatile by heat; the oxide breaking up into oxygen and mercury, which sublimes, while the sulphide sublimes unaltered, unless previously mixed with sodium carbonate or some reducing agent.

II. BISMUTH (Bi).

(*a*) *WET REACTIONS.*

(To be practised upon *bismuth subnitrate*, dissolved in water by the aid of the smallest possible quantity of nitric acid, and any excess of the latter carefully boiled off.)

1. **H_2S** *after acidulation by* **HCl** *(2nd group reagent)* gives a black precipitate of bismuth sulphide—Bi_2S_3—insoluble in ammonium sulphide, but soluble in boiling nitric acid.
2. **H_2SO_4** gives no precipitate (distinction from lead).
3. **NH_4HO, KHO**, and **NaHO**, all give precipitates of white bismuthous hydrate —$Bi_2H_2O_4$ or BiO(OH)—insoluble in excess and becoming converted into the yellow oxide—Bi_2O_3—on boiling.
4. **H_2O** in excess to a solution in which the free acid has been as much as possible driven off by boiling, gives a white precipitate of a basic salt of bismuth. This reaction is more delicate in the presence of hydrochloric than of nitric acid; and the precipitate, which is in this case bismuth oxy-chloride—BiOCl—is insoluble in tartaric acid (distinction from antimonious oxy-chloride).
5. **K_2CrO_4** yields a yellow precipitate of bismuth oxy-chromate—$Bi_2O_2CrO_4$—soluble in dilute nitric acid, but not in potassium hydrate (distinction from plumbic chromate).
6. **KI** gives brown bismuthous iodide, soluble in excess.
7. **Alkaline Carbonates** give white precipitates of bismuth oxy-carbonate, insoluble in excess.
8. Fragments of zinc added to a solution of bismuth cause a deposit of the metal as a dark grey powder.

(*b*) *DRY REACTION.*

(To be practised upon *bismuth subnitrate.*)

Mixed with sodium carbonate and heated on charcoal before the blowpipe, a hard bead of metallic bismuth is produced, and the surrounding charcoal is incrusted with a coating of oxide, deep orange-yellow while hot and pale yellow on cooling.

III. COPPER (Cu).

(*a*) *WET REACTIONS.*

(To be practised with a solution of cupric sulphate—$CuSO_4$.)

1. **H_2S** *after acidulation with* **HCl** *(2nd group reagent)* forms a precipitate of brownish-black cupric sulphide—CuS—which is nearly insoluble in ammonium sulphide but soluble in nitric acid. Its precipitation is prevented by the presence of potassium cyanide (distinction from cadmium). When long exposed to the air in a moist state, it oxidises to cupric sulphate and dissolves spontaneously.
2. **NH_4HO** causes a pale blue precipitate instantly soluble in excess, forming a deep blue solution of tetrammonio-cupric sulphate—$(NH_3)_4CuSO_4H_2O$.
3. **K_4FeCy_6** yields a chocolate-brown precipitate of cupric ferrocyanide—Cu_2FeCy_6. This test is very delicate, and is not affected by the presence of a dilute acid, but does not take place in an alkaline liquid.
4. **KHO** or **NaHO** precipitates light-blue cupric hydrate—$Cu(HO)_2$—insoluble in excess, but turning to black cupric oxy-hydrate—$(CuO)_2Cu(HO)_2$—on boiling.

5. **$KNaC_4H_4O_6$** and **NaHO** added successively, the latter in excess, produce a deep blue liquid (Fehling's solution), which when boiled with a solution of glucose (grape sugar) deposits brick-red cuprous oxide—Cu_2O.
6. The **Alkaline Carbonates** behave like their respective hydrates.
7. Fragments of zinc or iron precipitate metallic copper from solutions acidulated with HCl.

(*b*) *DRY REACTIONS.*

(To be practised upon cupric oxide—CuO—or *verdigris*—$Cu_2O(C_2H_3O_2)_2$.)

1. Heated with Na_2CO_3 and KCy on charcoal, in the inner blowpipe flame, red scales of copper are formed.
2. Heated in the borax bead before the outer blowpipe flame, colors it green while hot and blue on cooling. By carefully moistening the bead with $SnCl_2$ and again heating, this time in the inner flame, a red color is produced.

IV. CADMIUM (Cd).

(*a*) *WET REACTIONS.*

(To be practised with a solution of cadmium iodide—CdI_2.)

1. **H_2S** *after acidulation with* **HCl** *(2nd group reagent)* gives a yellow precipitate of cadmium sulphide—CdS—insoluble in ammonium sulphide, but soluble in boiling nitric acid. This precipitate does not form readily in presence of much acid; but its production is not hindered by the addition of potassium cyanide (distinction from copper).
2. **NH_4HO** produces a white precipitate of cadmium hydrate—$Cd(HO)_2$—soluble in excess.
3. **KHO** or **NaHO** both give precipitates of cadmium hydrate—$Cd(HO)_2$—insoluble in excess (distinction from zinc).
4. **Alkaline Carbonates** precipitate cadmium carbonate—$CdCO_3$—insoluble in excess.

(*b*) *DRY REACTION.*

(To be practised on cadmium carbonate—$CdCO_3$.)

Heated on charcoal before the blowpipe, a brownish incrustation of oxide is produced, owing to reduction of the metal and its subsequent volatilisation and oxidation by the outer flame.

DIVISION B.

Metals which are precipitated by sulphuretted hydrogen in the presence of hydrochloric acid, but yield sulphides which are soluble in ammonium sulphide.

I. ARSENIC (As).

(*a*) *WET REACTIONS.*

(To be practised with a solution of arsenious anhydride in boiling water slightly acidulated by hydrochloric acid.)

1. **H_2S**, *after acidulation with* **HCl**, causes a yellow precipitate of arsenious sulphide—As_2S_3—soluble in ammonium sulphide, forming ammonium sulpharsenite—$(NH_4)_3AsS_3$—but insoluble in strong boiling hydrochloric acid (distinction from the sulphides of Sb and Sn). This pre-

cipitate is also soluble in cold solution of commercial carbonate of ammonia (distinction from the sulphides of Sb, Sn, Au, and Pt). Dried and heated in a small tube with a mixture of Na_2CO_3 and KCy, it yields a mirror of arsenic.

2. Boiled with **KHO** *and a fragment of* **Zinc**, arseniuretted hydrogen—AsH_3—is evolved, which stains black a paper moistened with solution of argentic nitrate and held over the mouth of the tube during the ebullition (*Fleitmann's test*).

3. Boiled with ¼ of its bulk of **HCl** *and a slip of* **Copper**, a grey coating is deposited on the copper of cupric arsenide. On drying the copper carefully, cutting it into fragments, and heating in a wide tube, a crystalline sublimate of arsenious anhydride—As_2O_3—is obtained, which, when dissolved in water, gives a yellow precipitate of argentic arsenite—Ag_3AsO_3—with solution of ammonio-nitrate of silver (*Reinch's test*).

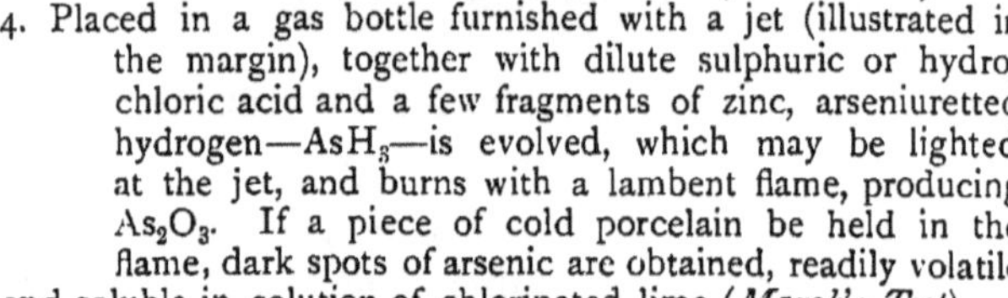

Fig. 16.

4. Placed in a gas bottle furnished with a jet (illustrated in the margin), together with dilute sulphuric or hydrochloric acid and a few fragments of zinc, arseniuretted hydrogen—AsH_3—is evolved, which may be lighted at the jet, and burns with a lambent flame, producing As_2O_3. If a piece of cold porcelain be held in the flame, dark spots of arsenic are obtained, readily volatile by heat and soluble in solution of chlorinated lime (*Marsh's Test*).

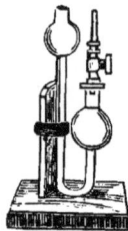

Note.—For reactions of **arsenites** and **arseniates**, see Acidulous Radicals.

(*b*) *DRY REACTION.*

(To be practised on arsenious anhydride—As_2O_3.)

Heated in a small tube with Na_2CO_3 and KCy, a mirror of arsenic is produced, accompanied by a garlic-like odor. The same effect may be produced with black flux.

II. ANTIMONY (Sb).

(*a*) *WET REACTIONS.*

(To be practised with a solution of tartar emetic.)

1. **H_2S**, *after acidulation by* **HCl**, causes an orange precipitate of antimonious sulphide—Sb_2S_3—soluble in ammonium sulphide, forming ammonium sulphantimonite—$(NH_4)_3SbS_3$—also soluble in strong boiling hydrochloric acid, forming antimonious chloride—$SbCl_3$—but insoluble in cold solution of commercial carbonate of ammonia.

2. **KHO** *and* **NaHO** produce precipitates of antimonious oxide readily soluble in excess to form antimonites (K_3SBO_3 or Na_3SbO_3).

3. Acidulated with **HCl** and introduced into a **platinum dish** with a rod of **zinc** so held that it touches the platinum *outside* the liquid, a black stain of metallic antimony is produced closely adherent to the platinum. This stain is not dissolved by HCl (tin reduced in the same manner is granular and soluble in boiling HCl).

4. **Reinch's test** (see Arsenic) produces a black coating on the copper, which, when heated, forms an amorphous sublimate of Sb_2O_3 *close to the copper*, and insoluble in water, but dissolved by a solution of cream of tartar in which H_2S then produces the characteristic orange sulphide.

5. **Marsh's test** (see Arsenic) yields stains of antimony on the porcelain, not

nearly so readily volatile by heat as in the case of arsenic, and not discharged by solution of chlorinated lime.

6. **Fleitmann's test** will not act with antimony at all (distinction from arsenic).

(*b*) *DRY REACTION.*

(To be practised on antimonious oxide—Sb_2O_3.)

Heated on charcoal with Na_2CO_3 and KCy before the blowpipe, a bead of metallic antimony is formed and copious white fumes of the oxide are produced.

III. TIN (Sn^{II} or Sn^{IV}).

(*a*) *WET REACTIONS.*

(To be practised with a solution of stannous chloride—$SnCl_2$—and one of stannic chloride—$SnCl_4$—prepared by warming the stannous solution with a little nitric acid.)

1. **H_2S,** *after acidulation with* **HCl**, produces a brown or yellow precipitate of SnS or SnS_2 respectively, both soluble in ammonium sulphide and in boiling hydrochloric acid.
2. **KHO** or **NaHO** both produce white precipitates of $Sn(HO)_2$ or $Sn(HO)_4$, soluble in excess, the former to produce *stannites* and the latter *stannates.* The stannous solution is, however, reprecipitable on boiling, while the stannic is not.
3. **NH_4HO** produces similar precipitates, very difficultly soluble in excess.
4. Acidulated by **HCl**, and introduced into a platinum dish with a rod of **zinc**, so held in the fluid that it touches the platinum *outside* the liquid, granules of metallic tin are deposited, soluble in boiling HCl, to form stannous chloride.
5. **$HgCl_2$** boiled with *stannous* salts deposits a grey precipitate of metallic mercury.

(*b*) *DRY REACTION.*

(To be practised on *putty powder*—SnO_2.)

Heated on charcoal with Na_2CO_3 before the blowpipe, a bead of metallic tin is produced, and a white incrustation of oxide is formed on the charcoal.

IV. GOLD (Au).

(*a*) *WET REACTIONS.*

(To be practised with a solution of auric chloride—$AuCl_3$.)

1. **H_2S** (*group reagent*) in the presence of HCl gives black auric sulphide —Au_2S_3. If the solution be hot, aurous sulphide—Au_2S—falls. Both are only soluble in nitro-hydrochloric acid, but they are soluble in ammonium sulphide when it is yellow.
2. **NH_4HO** precipitates reddish ammonium aurate, or *fulminating gold*—$Au_2(NH_3)_2O_2$—but KHO gives no result.
3. **$H_2C_2O_4$** (or **$FeSO_4$**) when boiled with an acid solution throws down Au. Reducing agents generally act thus. The liquid containing the metal may exhibit a blue, green, purple, or brown color.
4. **$SnCl_2$** throws down a brownish or purplish precipitate, known as "purple of Cassius," consisting of the mixed oxides of gold and tin.
5. **Zn, Cu, Fe, Pt,** or almost any metal, gives a precipitate of metallic Au in a finely divided state.

(*b*) *DRY REACTION.*

(To be practised on any gold salt.)

Heated on charcoal with Na_2CO_3, the metal is produced.

V. PLATINUM (Pt).

(*a*) *WET REACTIONS.*

(To be tested with a solution of platinic chloride—$PtCl_4$.)

1. **H_2S** (*2nd group reagent*) in presence of HCl gives a *brown* precipitate of platinic sulphide—PtS_2. This precipitate forms slowly, and is readily dissolved by yellow ammonium sulphide.
2. **KCl** *in presence of* **HCl**, especially after addition of alcohol, produces a yellow crystalline precipitate of potassium platinic chloride—PtK_2Cl_6—soluble to a moderate extent in water, but not in alcohol. Decomposition takes place when this is strongly heated, metallic Pt and KCl remaining.
3. **NH_4Cl** gives a precipitate of ammonium platinic chloride—$Pt(NH_4)_2Cl_6$,—which is almost identical in properties, but is more readily decomposed by heat, pure platinum remaining.
4. **Zn**, **Fe**, and several other metals decompose platinic salts with the production of the metal.

(*b*) *DRY REACTION.*

(To be practised upon potassium platinic chloride—PtK_2Cl_6.)

Heat on charcoal, with or without Na_2CO_3, before the blowpipe. The metal is produced by reduction.

GROUP III.

Metals which escape precipitation by sulphuretted hydrogen in presence of hydrochloric acid, but which are precipitated by ammonium sulphide in the presence of ammonium hydrate, ammonium chloride having been previously added to prevent the precipitation of magnesium.

DIVISION A.

Metals which, *in the insured absence of organic matter*, are precipitated as hydrates by the addition of the ammonium chloride and ammonium hydrate only.

I. IRON (*Ferrous*, **Fe**; and *Ferric*, **Fe_2**).

(*a*) *WET REACTIONS.*

(To be practised successively on solutions of ferrous sulphate—$FeSO_4$—and ferric chloride—Fe_2Cl_6.)

1. **NH_4HO** *in the presence of* **NH_4Cl** (*group reagent*) yields either a dirty-green precipitate of ferrous hydrate—$Fe(HO)_2$—or a reddish-brown precipitate of ferric hydrate—$Fe_2(HO)_6$. The former is *slightly* soluble in excess, but the latter is insoluble, and it is therefore preferable always to warm the solution with a little nitric acid, to insure the raising of the iron to the ferric state, before adding the ammonium hydrate. The presence of organic acids, such as tartaric or citric,

prevents the occurrence of this reaction; and therefore, if any such admixture be suspected, the solution should first be evaporated to dryness, the residue heated to redness, and then dissolved in a little hydrochloric acid, heated with a drop or two of nitric acid, diluted, and lastly, the NH_4Cl and NH_4HO added and boiled.

2. **NH_4HS**, added to a neutral or alkaline solution, produces a precipitate of ferrous sulphide—FeS—which is black (distinction from Al, Ce, Cr, Mn, and Zn), and readily soluble in cold diluted hydrochloric acid (distinction from the black sulphides of Ni and Co). This reaction takes place even in the presence of organic matter, and the precipitated sulphide, if exposed to the air, gradually oxidises to ferrous sulphate—$FeSO_4$—and disappears. It is insoluble in acetic acid (distinction from MnS).
3. **K_4FeCy_6**, in a neutral or slightly acid solution, gives, with ferrous salts, a white precipitate (rapidly changing to pale blue) of Everett's salt—potassium ferrous ferrocyanide—$K_2Fe \cdot FeCy_6$—and with ferric salts, a dark blue precipitate of Prussian blue—ferric ferrocyanide $(Fe_2)_2(FeCy_6)_3$. These precipitates are decomposed by alkalies, producing the hydrates of iron, and forming a ferrocyanide of the alkali in solution; but the addition of hydrochloric acid causes the re-formation of the original precipitate.
4. **$K_6Fe_2Cy_{12}$** gives, with ferrous salts, in neutral or slightly acid solutions, a dark blue precipitate of Turnbull's blue—ferrous ferricyanide, $Fe_3Fe_2Cy_{12}$—but with ferric salts it gives no precipitate, simply producing a brownish liquid. With alkalies, Turnbull's blue is decomposed, yielding black ferroso-ferric hydrate, and a ferricyanide of the alkali; but the addition of hydrochloric acid reproduces the original blue.
5. **KCyS** gives no precipitate with ferrous salts, but with ferric compounds it yields a deep blood-red solution. This color is not discharged by dilute hydrochloric acid (distinction from ferric acetate), but immediately bleached by solution of mercuric chloride (distinction from ferric meconate).
6. **KHO**, or **NaHO**, produces effects similar to those of ammonium hydrate.
7. **Na_2HPO_4** *in the presence of* **$NaC_2H_3O_2$** *or* **$NH_4C_2H_3O_2$** gives a whitish gelatinous precipitate of ferrous or ferric phosphates—$Fe_3(PO_4)_2$ or $Fe_2(PO_4)_2$—insoluble in acetic acid, but soluble in hydrochloric acid. The previous addition of citric or tartaric acids prevents this reaction.
8. **$NaC_2H_3O_2$**, added in excess to *ferric* salts, produces a deep red solution of ferric acetate—$Fe_2(C_2H_3O_2)_6$—which on boiling deposits as a reddish-brown ferric oxyacetate—$Fe_2O(C_2H_3O_2)_4$. This precipitate dissolves slightly on cooling; but iron can be entirely precipitated in this form if the solution be instantly filtered while hot.
9. **Alkaline Carbonates**, added to a ferrous salt, precipitate white ferrous carbonate—$FeCO_3$—but with ferric salts, throw down the reddish-brown ferric hydrate already described.

(*b*) *DRY REACTIONS.*

(To be practised on ferric oxide.)

1. Heated on charcoal before the inner blowpipe flame, a black magnetic powder is obtained, which is not the metal, but is ferroso-ferric oxide —Fe_3O_4.
2. Heated in the borax bead in the inner blowpipe flame, a bottle-green color is produced; but in the outer flame the bead is deep red while hot, and very pale yellow when cold.

II. CERIUM (Ce).

(a) WET REACTIONS.

(To be practised on cerous chloride—$CeCl_2$—prepared by boiling cerum oxalate with sodium hydrate, washing the insoluble cerous hydrate with boiling water, and dissolving it in the least possible excess of hydrochloric acid.)

1. **NH_4HO** *in the presence of* **NH_4Cl** (*group reagent*) gives a white precipitate of cerous hydrate—$Ce(HO)_2$—insoluble in excess.
2. **KHO** and **NaHO** give a similar precipitate, turning to yellow ceroso-ceric oxide—Ce_3O_4—on the addition of chlorine water.
3. **$(NH_4)_2C_2O_4$** gives a white precipitate of cerous oxalate—CeC_2O_4—insoluble in excess, and not readily dissolved even by hydrochloric acid. The presence of citric or tartaric acid does not interfere with this reaction.
4. **K_2SO_4** in a saturated solution causes the formation of white crystalline potassium cerium sulphate—$K_2Ce(SO_4)_2$—soluble in hot water.

(b) DRY REACTIONS.

(To be practised on cerium oxalate.)

1. Heated to redness in contact with the air, a deep red residue of ceric oxide—Ce_2O_3—is obtained, difficultly soluble even in strong hydrochloric acid.
2. Heated in the borax bead, cerium behaves like iron in the outer flame, but the inner flame yields a colorless or opaque yellow bead.

III ALUMINIUM (Al).

(a) WET REACTIONS.

(To be practised on a solution of *common alum.*)

1. **NH_4HO** *in presence of* **NH_4Cl** (*group reagent*) gives a gelatinous white precipitate of aluminic hydrate—$Al_2(HO)_6$. This precipitate is slightly soluble in a large excess of the precipitant, but separates completely on boiling.
2. **KHO** and **NaHO** both give a similar precipitate, soluble in excess, but reprecipitated by boiling with an excess of ammonium chloride, or by neutralising with hydrochloric acid and boiling with a slight excess of ammonium hydrate.
3. **Na_2HPO_4** *in the presence of* $NaC_2H_3O_2$ or $NH_4C_2H_3O_2$ gives a white precipitate of aluminic phosphate—$AlPO_3$—insoluble in hot acetic acid, but soluble in hydrochloric acid. The presence of citric or tartaric acids prevents the occurrence of this reaction.

(b) DRY REACTION.

(To be practised on dried *alum.*)

Heat strongly on charcoal before the blowpipe, when a strong incandescence is observed, and a white residue is left. Moisten this residue with a drop of solution of cobaltous nitrate—$Co(NO_3)_2$—and again heat strongly, when a blue mass is left. This test is not decisively characteristic, as other substances, such as zinc and earthy phosphates, show somewhat similar colours.

IV. CHROMIUM (Cr).

(*a*) *WET REACTIONS.*

(To be practised on a solution of potassium chromic chloride, prepared by dissolving potassium dichromate—$K_2Cr_2O_7$—in water, acidulating with hydrochloric acid, heating and dropping in rectified spirit till the solution turns green.)

1. **NH_4HO** *in the presence of* **NH_4Cl** (*group reagent*) precipitates green chromic hydrate—$Cr_2(HO)_6$—slightly soluble in excess, but entirely reprecipitated on boiling. The presence of citric or tartaric acid interferes with the completeness of this reaction.
2. **KHO**, or **NaHO**, gives similar precipitates, freely soluble in excess when cold, but entirely reprecipitable by continued boiling.
3. **NaOCl**, or **PbO_2**, boiled with an alkaline solution of a chromium salt, produçes a yellow solution of sodium chromate—Na_2CrO_4.
4. **$NaHPO_4$** *in the presence of* **$NaC_2H_3O_2$** or **$NH_4C_2H_3O_2$** throws down pale green chromic phosphate—$CrPO_4$—soluble when freshly precipitated in excess of hot acetic acid, and freely soluble in hydrochloric acid. The presence of organic acids prevents this reaction.

(*b*) *DRY REACTIONS.*

1. Heated in the borax bead in the inner blowpipe flame, a fine green color is obtained.
2. Fused on platinum foil, with a mixture of $KNaCO_3$ and KNO_3, a yellow residue is obtained, consisting of chromates of the alkalies used. This mass is soluble in water, yielding a yellow solution turned deeper in color by the addition of hydrochloric acid, owing to the formation of dichromates, and becoming green on warming and dropping in rectified spirit.

DIVISION B.

Metals the hydrates of which, being soluble in excess of ammonium hydrate in the presence of ammonium chloride, escape precipitation by that reagent, but are separated as insoluble sulphides by the addition of ammonium sulphide to the same liquid.

I. MANGANESE (Mn).

(*a*) *WET REACTIONS.*

(To be practised on a solution of potassium manganous chloride, prepared by heating a solution of potassium permanganate with hydrochloric acid, and dropping in rectified spirit until a colorless solution is obtained.)

1. **NH_4HS** *in the presence of* **NH_4Cl** *and* **NH_4HO** (*group reagent*) precipitates a flesh-colored manganous sulphide—MnS—soluble in dilute and cold hydrochloric acid (distinction from the sulphides of Ni and Co). It is also soluble in acetic acid (distinction from zinc sulphide). This precipitate forms sometimes very slowly and only after gently warming. If a good excess of NH_4Cl has not been added, or if, after adding the excess of ammonium hydrate, the solution be exposed to the air, a portion of the manganese will sometimes precipitate spontaneously, as manganic dioxyhydrate—$Mn_2O_2(HO)_2$—and be found with the iron, etc., in the first division of the third group. In this case its

presence will be easily made manifest during the fusion for chromium by the residue being green. It is therefore evident that small quantities of manganese cannot be perfectly separated from large quantities of iron by NH_4Cl and NH_4HO only.

2. **KHO** and **NaHO** both yield precipitates of manganous hydrate insoluble in excess, and converted by boiling into dark brown manganic dioxyhydrate—$Mn_2O_2(HO)_2$.
3. **NH_4HO** gives a similar precipitate, soluble in excess of ammonium chloride, but gradually depositing as $Mn_2O_2(HO)_2$ by exposure to the air. For this reason, if the presence of manganese be suspected, the addition of NH_4Cl and NH_4HO must be followed by instant filtration, and any cloudiness coming in the filtrate must be simply taken as indicating manganese, and disregarded.
4. **K_4FeCy_6** gives a precipitate of manganous ferrocyanide—Mn_2FeCy_6—very liable to be mistaken for the corresponding zinc compound.
5. Boiled with plumbic peroxide and nitric acid a violet color is produced in the liquid, due to the formation of permanganic acid. (*Crum's test.*)

(*b*) *DRY REACTIONS.*

(To be practised upon manganese peroxide—MnO_2.)

1. Fused on platinum foil with $KNaCO_3$ and KNO_3, a green mass of manganates of the alkalies is formed. This residue is soluble in water, yielding a green solution, turning purple on boiling, owing to the formation of permanganates. The solution is rendered colorless by heating with hydrochloric acid and dropping in rectified spirit, the operation being accompanied by the odor of aldehyd.
2. Heated in the borax bead in the outer blowpipe flame, a color is produced which is violet-red while hot and amethyst on cooling. The bead is rendered colorless by the reducing flame.

II. ZINC (Zn).

(*a*) *WET REACTIONS.*

(To be practised on zinc sulphate—$ZnSO_4$.)

1. **NH_4HS** *in the presence of* **NH_4Cl** *and* **NH_4HO** (*group reagent*) gives a white precipitate of zinc sulphide—ZnS—insoluble in acetic acid, but readily soluble in dilute hydrochloric acid.
2. **KHO**, **NaHO**, and **NH_4HO**, all give precipitates of gelatinous white zinc hydrate, soluble in excess to form zincates. The addition of sulphuretted hydrogen or ammonium sulphide reprecipitates the zinc as zinc sulphide—ZnS.
3. **K_4FeCy_6** gives a gelatinous white precipitate of zinc ferrocyanide—Zn_2FeCy_6—insoluble in dilute acids.
4. **Alkaline Carbonates** precipitate $ZnCO_3(Zn2HO)_2H_2O$—zinc hydratocarbonate—insoluble in excess of the carbonates of potassium and sodium, but soluble in that of ammonium. The latter solution, diluted and boiled, deposits the oxide.

(*b*) *DRY REACTIONS.*

(To be practised on zinc carbonate.)

1. Salts of zinc heated leave the oxide, yellow while hot, and white on cooling.

2. Heated on charcoal before the blowpipe an incrustation forms, yellow while hot, and white on cooling. Moisten with a drop of cobaltous nitrate—$Co(NO_3)_2$—and again heat in the outer flame, when a fine green color is produced.

III. NICKEL (Ni).

(*a*) *WET REACTIONS.*

(To be practised on a solution of nickelous sulphate—$NiSO_4$.)

1. **NH_4HS** *in the presence of* **NH_4Cl** *and* **NH_4HO** (*group reagent*) gives a black precipitate of nickelous sulphide—NiS—slightly soluble in excess, but entirely precipitated on boiling. It is not soluble in cold dilute hydrochloric or in acetic acid, but requires boiling with strong hydrochloric acid, and sometimes even the addition of a drop or two of nitric acid.
2. **KHO** or **NaHO** both give a green precipitate of nickelous hydrate—$Ni(HO)_2$—unaltered by boiling (distinction from cobalt).
3. **KNO_2** added to a neutral solution, followed by an excess of acetic acid, gives no precipitate (after standing some hours) on the addition of potassium acetate and rectified spirit (very useful separation from cobalt).
4. **KCy** in excess produces a greenish-yellow precipitate of nickelous cyanide —$NiCy_2$—which quickly redissolves. On adding a drop of hydrochloric acid and boiling in a fume chamber, and repeating this till no more fumes of hydrocyanic acid come off, and then adding sodium hydrate, a precipitate of nickel hydrate is produced. It is better, although less convenient, to use a strong solution of chlorinated soda —NaClNaClO—in the last stage, when nickelic hydrate—$Ni(HO)_6$—is slowly precipitated (separation from Co, which gives no precipitate).
5. **Alkaline Carbonates** behave, so far as color and solubility in excess are concerned, like their respective hydrates.

(*b*) *DRY REACTIONS.*

1. Heated on charcoal with Na_2CO_3 in the inner blowpipe flame, a grey metallic and magnetic powder is produced.
2. Heated in the borax bead in the outer blowpipe flame, red to violet-brown is produced while hot, and a yellowish to sherry-red when cold These colors might be mistaken for those of iron; but on fusing a small fragment of potassium nitrate with the bead, its color at once changes to blue or dark purple (distinction from Fe).

IV. COBALT (Co).

(*a*) *WET REACTIONS.*

(To be practised on a solution of cobaltous nitrate—$Co(NO_3)_2$.)

1. **NH_4HS** *in the presence of* **NH_4Cl** *and* **NH_4HO** (*group reagent*) gives a black precipitate of cobaltous sulphide—CoS—insoluble in acetic and cold dilute hydrochloric acid, and requiring to be boiled with the strongest HCl, often with the addition of a drop or two of nitric acid before solution is effected.
2. **KHO**, or **NaHO**, gives a blue precipitate of cobaltous hydrate—$Co(HO)_2$—rapidly changing to pink on boiling (distinction from nickel).
3. **KCy** gives a light brown precipitate of cobaltous cyanide, rapidly soluble

in excess but reprecipitated by excess of dilute hydrochloric acid. If, however, the HCl be added drop by drop just so long as it causes the evolution of hydrocyanic acid fumes on boiling,* soluble potassium cobalticyanide—$K_6Co_2Cy_{12}$—results, which is not decomposed by hydrochloric acid; nor is any precipitate produced on adding excess of sodium hydrate or chlorinated soda (separation from nickel).

4. **Alkaline Carbonates** throw down basic carbonates, behaving like the respective hydrates.

(*b*) *DRY REACTIONS.*

1. Heated on charcoal with Na_2CO_3 in the inner blowpipe flame, the cobalt separates as a grey magnetic powder.
2. Heated in the borax bead, first in the *outer* and then in the *inner* flame, a fine blue color is produced. It is an important distinction of cobalt from copper, manganese, etc., that *prolonged heating in the inner flame does not affect this blue.*

GROUP IV.

Metals the hydrates and sulphides of which, being soluble, are not precipitated by the addition of NH_4HO and NH_4HS in the presence of NH_4Cl, but separate as insoluble carbonates on the addition of ammonium carbonate to the same solution.

I. BARIUM (Ba).

(*a*) *WET REACTIONS.*

(To be practised on a solution of barium chloride—$BaCl_2$.)

1. **$(NH_4)_2CO_3$** *in the presence of* NH_4Cl *and* NH_4HO (*group reagent*) produces a white precipitate of barium carbonate, $BaCO_3$, soluble with effervescence in dilute acetic acid.
2. **H_2SO_4** and all **soluble Sulphates** give a white precipitate of barium sulphate—$BaSO_4$—insoluble in ammonium acetate or tartrate (distinction from $PbSO_4$) and also in boiling nitric acid.
3. **K_2CrO_4** gives a yellow precipitate of barium chromate—$BaCrO_4$—insoluble in water and in dilute acetic acid, but soluble in hydrochloric acid (distinction from Sr and Ca).
4. **$(NH_4)_2C_2O_4$** gives a white precipitate of barium oxalate—BaC_2O_4—not readily formed in the presence of much acetic acid.
5. **Na_2HPO_4** gives a white precipitate of barium hydrogen phosphate—$BaHPO_4$—soluble in acetic acid, and to some extent in ammonium chloride.

(*b*) *DRY REACTION.*

(To be also practised on barium chloride.)

If a platinum wire be dipped first in hydrochloric acid and then in the salt and held in the *inner* blowpipe or *Bunsen* flame, the outer flame is colored yellowish-green.

---

* This must be done in a fume chamber, as it is a highly poisonous operation if the fumes should happen to escape into the room.

II. STRONTIUM (Sr).

(*a*) *WET REACTIONS.*

(To be practised on strontium nitrate—$Sr(NO_3)_2$.)

1. **$(NH_4)_2CO_3$** (*group reagent*) *in the presence of* **NH_4Cl** *and* **NH_4HO** gives a white precipitate of strontium carbonate—$SrCO_3$—soluble in dilute acetic acid.
2. **H_2SO_4**, or a soluble sulphate (preferably calcium sulphate), yields a white precipitate of strontium sulphate—$SrSO_4$—which only separates completely from dilute solutions on allowing them to stand in a warm place for some hours. It is insoluble in a boiling strong solution of ammonium sulphate rendered alkaline by ammonium hydrate (distinction from calcium sulphate).
3. The other reactions are similar to those of calcium.

(*b*) *DRY REACTION.*

(To be also practised on $Sr(NO_3)_2$.)

A platinum wire moistened with hydrochloric acid, dipped in the substance and introduced into the *inner* blowpipe or *Bunsen* flame, colors the outer flame crimson.

III. CALCIUM (Ca).

(*a*) *WET REACTIONS.*

(To be practised on a solution of calcium chloride—$CaCl_2$.)

1. **$(NH_4)_2CO_3$** *in presence of* **NH_4Cl** *and* **NH_4HO** (*group reagent*) produces a white precipitate of calcium carbonate—$CaCO_3$—soluble in acetic acid and settling best on warming.
2. **$(NH_4)_2C_2O_4$** precipitates white calcium oxalate—CaC_2O_4—insoluble in acetic or oxalic acids, but soluble in hydrochloric acid.
3. **H_2SO_4** in strong solutions produces a precipitate of calcium sulphate—$CaSO_4$. Being slightly soluble in water, it does not form in dilute solutions, nor is it precipitated by a saturated solution of calcium sulphate (distinction from Ba and Sr). It is soluble in a boiling saturated solution of ammonium sulphate containing excess of ammonium hydrate, but quite insoluble in a mixture of two parts alcohol and one part water.
4. **Na_2HPO_4** produces a white precipitate of calcium phosphate soluble in acetic acid.

(*b*) *DRY REACTION*

(To be practised on calcium carbonate—$CaCO_3$.)

A platinum wire moistened with hydrochloric acid, dipped in the substance and held in the *inner* blowpipe or *Bunsen* flame, colors the outer flame yellowish-red. This reaction is masked by the presence of barium or strontium.

GROUP V.

Metals not precipitable either as sulphide, hydrate, or carbonate, including magnesium the precipitation of which as hydrate or carbonate has been prevented by the presence of ammonium chloride.

I. MAGNESIUM (Mg).

(*a*) *WET REACTIONS.*

(To be practised on a solution of magnesium sulphate—$MgSO_4$.)

1. **Na_2HPO_4** *in the presence of* NH_4Cl *and* NH_4HO produces a white crystalline precipitate of ammonium magnesium phosphate—$MgNH_4PO_4$. It is slightly soluble in water, and scarcely at all in water containing ammonium hydrate, but entirely soluble in all acids. In very dilute solutions it only forms on cooling and shaking violently, or on rubbing the inside of the tube with a glass rod.
2. **$(NH_4)_2HAsO_4$** produces a similar precipitate of $MgNH_4AsO_4$, possessing like features.
3. **KHO, NaHO**, and **NH_4HO** give precipitates of magnesium hydrate—$Mg(HO)_2$—insoluble in excess, but soluble in the presence of ammonium salts. The alkaline carbonates (except ammonium carbonate) precipitate magnesium carbonate, also soluble in ammonium salts.
4. **$Ca(HO)_2$** and **$Ba(HO)_2$** produce a similar effect. Either of these reagents is useful for the separation of magnesium from all the alkalies except ammonium. The solution, which must contain no ammonium salts, is treated with excess of either lime or baryta water. The precipitated magnesium hydrate is then filtered out and excess of ammonium carbonate added, which precipitates in turn the excess of Ca or Ba employed, and leaves K, Na, or Li in solution.

(*b*) *DRY REACTION.*

(To be practised on magnesium oxide.)

Heated on charcoal before the blowpipe, it becomes strongly incandescent, and leaves a white residue, which when moistened with a drop of solution of cobaltous nitrate—$Co(NO_3)_2$—and again heated, becomes rose-colored. This test is not, however, infallible.

II. LITHIUM (Li).

(*a*) *WET REACTIONS.*

(To be practised on a solution of lithium chloride, prepared by dissolving lithium carbonate in dilute hydrochloric acid.)

1. **Na_2HPO_4** in strong solutions produces a white precipitate of lithium phosphate—Li_3PO_4—on boiling only (distinction from Mg). It is soluble in hydrochloric acid, and reprecipitated by boiling with ammonium hydrate.
2. **Na_2CO_3** and even NaHO, in very strong solutions, yield the carbonate and hydrate respectively.
3. **$PtCl_4$** gives no precipitate (distinction from potassium).

(*b*) *DRY REACTION.*

(To be practised with lithium carbonate.)

A platinum wire, moistened with hydrochloric acid, dipped in the substance and held in the *inner* blowpipe or *Bunsen* flame, colors the outer flame *carmine red.* The presence of sodium disguises this reaction.

III. POTASSIUM (K).

WET REACTIONS.

(To be practised on solution of potassium carbonate treated with dilute HCl till effervescence ceases, forming potassium chloride—KCl.)

1. **$PtCl_4$**, in strong solutions, gives a yellow crystalline precipitate of potassium platino-chloride—K_2PtCl_6—soluble on great dilution, especially on warming, but insoluble in acids, alcohol and ether.
2. **$H_2C_4H_4O_6$** throws down, from strong solutions only, a white crystalline precipitate of potassium hydrogen tartrate—$KHC_4H_4O_6$—soluble in much cold water, rather freely in hot water, readily in acids and in KHO or NaHO, and not formed unless the original solution be nearly neutral. Its separation is facilitated by stirring and shaking violently, in which case it settles quickly.
3. **H_2SiF_6** (*hydrofluosilicic acid*) yields white gelatinous potassium fluosilicate—K_2SiF_6—sparingly soluble in water.

DRY REACTION.

(To be practised on potassium carbonate—K_2CO_3.)

Dip a platinum wire, moistened with HCl, in the salt. Held in a *Bunsen* flame a *violet color* is imparted. The masking effect of Na (*yellow*) is obviated by viewing the flame through indigo glass.

IV. SODIUM (Na).

WET REACTIONS.

(To be tested with solution of sodium chloride—NaCl.)

1. **$KSbO_3$** (*potassium metantimoniate*) gives a white granular precipitate of *sodium metantimoniate*—$NaSbO_3$—from strong solutions only, which must be neutral or alkaline. This precipitate is insoluble in alcohol.
2. **H_2SiF_6** gives a similar precipitate to that obtained with K salts in concentrated solutions only.

Sodium salts are, practically, all soluble in water, and there is no thoroughly trustworthy wet reaction which can be applied to detect small quantities. If we have a solution which gives no precipitate with any of the group reagents, but leaves, on evaporating, a fixed residue capable of imparting a *strong yellow color* to the *Bunsen* flame (**dry reaction**), we may infer with certainty the presence of sodium.

V. AMMONIUM (NH_4).

WET REACTIONS.

(To be tested with solution of ammonium chloride—NH_4Cl.)

1. **$PtCl_4$** produces a heavy yellow precipitate of ammonium platino-chloride—$(NH_4)_2PtCl_6$—which, being rather soluble in water, is not formed in

dilute solutions, unless alcohol, in which it is insoluble, be added in considerable quantity. When ignited, pure spongy platinum is left. This precipitate may be distinguished from that with K salts by adding, after ignition, a little water and $AgNO_3$, when no white precipitate of AgCl is formed (the K salt leaves KCl on being strongly heated).

2. **$H_2C_4H_4O_6$** yields ammonium hydrogen tartrate—$(NH_4)HC_4H_4O_6$—almost identical with $KHC_4H_4O_6$ in its properties. On ignition, however, the latter gives a black residue, which turns moistened red litmus paper *blue* (K_2CO_3 and C), the former leaving pure C without reaction.
3. **NaHO** or **$Ca(HO)_2$** boiled with the solution causes the evolution of ammonia gas—NH_3. A glass rod dipped in HCl or $HC_2H_3O_2$ produces, when held over a mixture evolving NH_3, white clouds (solid NH_4 salts), and moist red litmus paper is turned blue.
4. **Nessler's Solution** (HgI_2 dissolved in KI and KHO added) gives a *yellow* or *brown color* or a *brown precipitate* of $N(Hg'')_2I$ with all NH_4 salts. This reaction is extremely delicate, and the estimation of NH_4 in water is founded upon it.

DRY REACTIONS.

Ammonium salts volatilise (1) with decomposition, leaving a fixed acid (*e.g.*, phosphate); (2) with decomposition, leaving no residue whatever (*e.g.*, sulphate, nitrate); (3) without decomposition, when they are said to *sublime* (*e.g.*, chloride, bromide, etc.)

CHAPTER III.

DETECTION AND SEPARATION OF ACIDULOUS RADICALS.

1. HYDROFLUORIC ACID and FLUORIDES.

(The test for fluorides undernoted may be practised on fluor spar—CaF_2.)

Hydrofluoric Acid, or **Fluoric Acid**, is known,—

1. By its strongly acid reaction and corrosive power.
2. By its action upon glass, from which it dissolves out silicic acid—SiO_2—thus roughening the surface and rendering it semi-opaque or translucent, and white; a colorless gas, silicic fluoride—SiF_4—passing off.

Fluorides are detected as follows:—

The mineral or salt is finely powdered, and introduced into a leaden dish with a little sulphuric acid. A piece of glass, previously prepared by coating its surface with wax, and etching a few letters on the waxed side with the point of a pin, is placed over the dish, waxed side down. A gentle heat is then applied, but not sufficient to melt the wax, and the operation continued for some time. The glass is then taken off, and the wax removed from it; when, if fluorine was present, the letters written on the waxed surface will be found engraved upon it by the action of the hydrofluoric acid.

2. CHLORINE, HYDROCHLORIC ACID, and CHLORIDES.

Free Chlorine—Cl_2—may be detected,—

1. By its odor.
2. By turning paper dipped in solution of potassium iodide brown.
3. By bleaching a solution of indigo or litmus.

Hydrochloric Acid—HCl—may be recognised,—

1. By its acidity and odor of its fumes.
2. By producing dense white fumes when a rod dipped in ammonium hydrate is held over the mouth of the bottle.
3. By giving a curdy white precipitate of argentic chloride with argentic nitrate, instantly soluble in ammonium hydrate.

Chlorides give the following reactions (to be practised with any soluble chloride, say NaCl):—

1. Heated with sulphuric acid they evolve white fumes of HCl.
2. Heated with $\mathbf{H_2SO_4}$ and $\mathbf{MnO_2}$ they evolve chlorine.
3. $\mathbf{AgNO_3}$ *in the presence of* HNO_3 gives a white precipitate of argentic chloride—AgCl—insoluble in boiling nitric acid, but instantly soluble in dilute ammonium hydrate of a strength of 1 in 20.

4. The solid substance mixed with $K_2Cr_2O_7$ and distilled with H_2SO_4 yields chloro-chromic acid—$CrCl_2O_2$—in red fumes which, when passed into dilute ammonium hydrate, color it yellow, owing to the formation of ammonium chromate—$(NH_4)_2CrO_4$. The yellow should change readily to green on the addition of a few drops of sulphurous acid.

Insoluble Chlorides should be first boiled with strong sodium hydrate and the whole diluted and filtered. The chloride is then transferred to the sodium, and is to be searched for in the filtrate by acidulating with nitric acid and adding argentic nitrate, as above described.

3. HYPOCHLORITES.

(Practise on a solution of chlorinated lime.)

Hypochlorites are all readily soluble in water, are contained in the so-called *chlorinated* compounds, and are recognised—

1. By having an odor of chlorine.
2. By giving a blue with potassium iodide, starch paste and acetic acid, due to liberation of chlorine.

4. CHLORATES.

(To be practised on potassium chlorate—$KClO_3$.)

1. Heated on charcoal, they deflagrate.
2. Heated with strong sulphuric acid, they evolve chlorine peroxide—Cl_2O_4—which is yellow and explosive.
3. Their solutions yield no precipitate with argentic nitrate; but if a little of the solid be heated to redness, and the residue dissolved in water, a precipitate of argentic chloride is obtained. The same reduction from chlorate to chloride may also be effected by adding zinc and dilute sulphuric acid to the solution.
4. Mixed with KI and starch paste and acidulated with acetic acid they give no blue (distinction from hypochlorites), but on adding HCl, a blue is developed.

5. SEPARATION OF CHLORATES AND CHLORIDES.

(Practise on mixed solutions of KCl and $KClO_3$.)

Add *excess* of argentic nitrate, filter out the argentic chloride formed, and then acidulate with sulphuric acid, and drop in a fragment of zinc, when, if a chlorate be present, a second precipitate of argentic chloride will form.

6. PERCHLORATES.

These are distinguished from chlorates—

1. By giving off perchloric acid, when heated with sulphuric acid, without explosion or evolution of chlorine peroxide.
2. Like chlorates, they require reduction to chlorides before giving a precipitate with argentic nitrate.

7. BROMINE, HYDROBROMIC ACID, and BROMIDES.

Bromine is distinguished,—

1. By its appearance—heavy, reddish-brown liquid, giving off reddish fumes of a very penetrating, unpleasant odor.

2. By turning starch paste yellow or pink.
3. When present in small quantity in solution, on adding a few drops of chloroform and shaking, an orange color is imparted to that liquid, which sinks to the bottom of the aqueous solution.

Hydrobromic Acid is known,—

By its acid reaction and the production of fumes of bromine when heated with strong sulphuric acid.

Bromides are all soluble in water, except the silver, mercurous and lead salts; they are detected by the following characters (to be practised on potassium bromide—KBr):—

1. Heated with **strong sulphuric acid**, they evolve red vapors of bromine.
2. A similar effect is produced by **sulphuric acid** and **metallic dioxides**, such as PbO_2, MnO_2.
3. Mixed with **starch paste**, and a few drops of **chlorine water** carefully added, they give an orange color (starch bromide).
4. Mixed in a long tube with **chloroform**, and a few drops of **chlorine water** added, the whole, when shaken well together, leaves a characteristic reddish-colored stratum at the bottom of the liquid in the tube, due to free bromine in the chloroform.
5. With **argentic nitrate** they give a dirty-white precipitate of **argentic bromide**, insoluble in nitric acid, slowly soluble in ammonium hydrate, but insoluble in dilute NH_4HO, of a strength of 1 in 20 (argentic chloride dissolves).
6. Distilled with **potassium dichromate** and **sulphuric acid**, red fumes are evolved, which give no color when passed into **ammonium hydrate** (distinction from chlorides).

Insoluble Bromides should be first boiled with NaHO, as described under insoluble chlorides.

8. DETECTION OF CHLORIDES IN THE PRESENCE OF BROMIDES.

(To be practised on a mixture of KCl and KBr.)

The solution is divided into two parts, in one of which the bromide is proved by the addition of chlorine water, and shaking up with chloroform. The second portion is either—(1) Evaporated to dryness, the residue placed in a tube retort with a little potassium dichromate and sulphuric acid, while into the receiver is placed a little dilute ammonium hydrate, and distillation is proceeded with, when, if a chloride be present, the liquid in the receiver will be colored yellow; or—(2) Precipitated with excess of $AgNO_3$, washed on a filter percolated with dilute NH_4HO (1 in 20) and nitric acid added to the percolate, when a *distinctly curdy* white precipitate proves the presence of chlorides. This latter method is simple, and rarely fails if, on adding the acid, a mere cloud be disregarded.

9. HYPOBROMITES.

These are very similar to hypochlorites, and react as follows:—

1. They decompose by heat, leaving a bromide;
2. On boiling with an alkali, a mixture of bromide and bromate results.

10. BROMATES.

These are recognised—

1. By deflagrating on charcoal, leaving the corresponding bromide.
2. By liberating bromine on the addition of dilute sulphurous acid.

11. IODINE, HYDRIODIC ACID, and IODIDES.

Iodine is readily known by its glistening black scales, its odor, the violet vapor on heating, and the production of blue iodide of starch on adding a solution to starch paste.

Hydriodic Acid, in the gaseous state, is known by the formation of a brown color on holding paper moistened with chlorine water (blue if also dipped in starch paste) over a tube in which it is being evolved.

Iodides are readily known by the following reactions (which may be practised on a solution of potassium iodide, KI):—

1. Heated with strong sulphuric acid they give a liberation of iodine with violet fumes.
2. **Mucilage of Starch** and **nitric acid** or **chlorine water** (if not added too plentifully), produces **blue** iodide of starch, decomposed by heat but re-formed on cooling; also destroyed by excess of Cl (iodine trichloride—ICl_3—being produced).
3. The light yellow precipitate of argentic iodide—AgI—formed when a solution (containing alkaline metals only) is added to **argentic nitrate** dissolved in water. The precipitate, when freed from the supernatant liquid, does not dissolve in hot HNO_3, and is practically insoluble in ammonium hydrate, being thus distinguished from a chloride.
4. A *neutral* solution (produced, if alkaline in the first place, by the cautious addition of dilute HNO_3; if acid, by dropping in KHO solution until test-paper is unaffected) gives with one part of **cupric sulphate**—$CuSO_4$—and three parts, or rather less, of **ferrous sulphate**—$FeSO_4$—dissolved in a little water, a greyish precipitate of **cuprous iodide**—$(Cu_2)''I'_2$.

   The same precipitate is produced if **sulphurous acid**—H_2SO_3—be used with the cupric sulphate instead of ferrous sulphate.
5. **Palladious Chloride**—$PdCl_2$—or **palladious nitrate**—$Pd(NO_3)_2$—gives a black precipitate of **palladious iodide**—PdI_2—decomposed somewhat below the temperature of boiling mercury, iodine being evolved, and the metal left. This is a very expensive but efficient separation.
6. Mercuric chloride and plumbic nitrate give respectively red and yellow precipitates with soluble iodides.

12. DETECTION OF BROMIDES IN THE PRESENCE OF IODIDES.

(Practise on a mixture of KBr and KI.)

Add to the solution a very small quantity of starch paste and then a drop or two of chlorine water, when a blue color will be produced, proving the iodide. Continue to add more chlorine water until this blue is entirely discharged, and then shake up with chloroform, when, if a bromide be present, the characteristic golden color will be communicated to the chloroform.

13. DETECTION OF CHLORIDES IN THE PRESENCE OF IODIDES.

(Practise on a mixture of KCl and KI.)

Add excess of argentic nitrate, warm, pour off the supernatant liquid, wash with warm water, and shake up the precipitate in dilute solution of ammonium hydrate (1 in 3). The argentic iodide will remain insoluble, while the chloride will dissolve and may be detected in the solution after filtration by reprecipitation with excess of nitric acid. As argentic iodide is not *absolutely* insoluble in ammonium hydrate, a mere cloud on adding the nitric acid is to be disregarded. This test is only accurate in the insured absence of a bromide, proved as above directed (see 12).

14. SEPARATION OF AN IODIDE FROM A BROMIDE AND CHLORIDE.

(Practise on a mixture of KCl, KBr, and KI.)

1. Add to the solution a mixture of one part **cupric sulphate** and three parts **ferrous sulphate**, or mix the solution with excess of **cupric sulphate** and drop in **sulphurous acid** till precipitation ceases. The iodide will separate as cuprous iodide—Cu_2I_2—leaving the bromide and chloride in solution. Unless carefully done, this separation is not absolutely accurate.
2. Add to the solution palladious nitrate until precipitation ceases. Filter out the palladious iodide which separates, and pass sulphuretted hydrogen through the liquid to remove excess of palladium, and again filter. Boil to expel the excess of H_2S, and the bromide and chloride remain in solution.

15. IODATES.

(Practise on solution of potassium iodate.)

Iodates are known,—

1. By giving, when heated with strong sulphuric acid, effects likely to be mistaken for chlorates.
2. By giving a blue with starch paste on the addition of sulphurous acid.
3. By giving a blue with starch paste on the addition of potassium iodide and tartaric acid.
4. By yielding a precipitate of ferric oxy-iodate on adding ferric chloride.

15a. DETECTION OF AN IODATE IN AN IODIDE.

(Practise on a solution of iodine in heated potassium hydrate—$KI+KIO_3$.)

When excess of tartaric acid is added to potassium iodate, iodic acid is set free; and when the same acid is added to potassium iodide, hydriodic acid is set free, and potassium hydro-tartrate formed. Thus:

$$5KI+KIO_3+6H_2C_4H_4O_6=5HI+HIO_3+6KHC_4H_4O_6.$$

If these acids be thus liberated together, they immediately decompose, forming water and free iodine:

$$5HI+HIO_3=3I_2+3H_2O.$$

If therefore starch paste and tartaric acid be added to pure potassium iodide no coloration takes place, because only hydriodic acid is liberated; but if the sample contains potassium iodate, an immediate production of free iodine ensues, which turns the starch blue.

16. PERIODATES.

Periodates are distinguished,—

1. By precipitating with **barium chloride**—$BaCl_2$—in a neutral solution and digesting in **ammonium carbonate** to which some NH_4HO has been added. The periodate is not decomposed. Iodates leave barium carbonate, which when washed dissolves in acid with effervescence.
2. By adding $Hg(NO_3)_2$ and treating the yellowish precipitate with $SnCl_2$. It turns green, HgI_2 being produced.

17. WATER and HYDRATES.

Water is recognised,—

1. By its absolute neutrality to test paper.
2. By its evaporating without residue, fumes or odor of any kind.

3

3. By its turning white anhydrous cupric sulphate blue.
4. By its yielding pure hydrogen when boiled and the steam passed slowly over copper turnings heated to bright redness in an iron tube.
5. By its undergoing electrolysis, and yielding hydrogen at the negative and oxygen at the positive electrode.

The **soluble Hydrates**, viz., KHO, NaHO, LiHO, $Ba(HO)_2$, $Sr(HO)_2$, and $Ca(HO)_2$ are known,—

1. By being more or less soluble in cold water, yielding solutions which are strongly alkaline to test-paper.
2. By dissolving in hydrochloric acid without effervescence and without smell.
3. By giving a brownish-black precipitate of argentic oxide—Ag_2O—with argentic nitrate.

The **insoluble Hydrates** are recognised,—

By giving off steam when heated in a dry test-tube, and leaving a residue which behaves like the corresponding oxide.

18. OXIDES.

All oxides are insoluble in water. Those of K, Na, Li, Ba, Sr, and Ca, when placed in contact with that liquid, unite with it to form hydrates, which dissolve with a greater or less degree of readiness and give the characters of the soluble hydrates already mentioned.

Normal Oxides can only be recognised by negative results, such as :—

1. Heated alone, they are not changed; except argentic oxide, which leaves the metal, and mercuric oxide, which volatilises and breaks up into the metal and oxygen.
2. They are insoluble in water (exceptions K, Na, Li, Ca, Ba, and Sr, converted into soluble hydrates), but soluble in hydrochloric or nitric acid without effervescence and without smell.
3. After dissolving and removing the metal by H_2S or Na_2CO_3 as most convenient, no acidulous radical is found, other than that of the acid used to dissolve.
4. Boiled with strong NaHO and filtered, or fused with $KNaCO_3$ and digested with water, the solution gives no reaction for any acid radical except the soluble hydrate or carbonate employed.

Peroxides, on account of their containing an excess of oxygen, differ from normal oxides (practise on MnO_2),—

1. By giving off oxygen when strongly heated.
2. By evolving chlorine when heated with hydrochloric acid.

19. SULPHUR, HYDROSULPHURIC ACID, and SULPHIDES.

Ordinary Sulphur is recognised,—

1. By its burning entirely away with a pale blue flame, and evolving sulphurous anhydride.
2. By its insolubility in all ordinary menstrua, such as water, alcohol, and ether, but dissolving readily in carbon disulphide.
3. When slowly heated in a tube, it first melts, then thickens, then melts again, and finally boils, the vapor taking fire and forming sulphurous anhydride.

Precipitated Sulphur possesses the above characters, and is specially distinguished from ordinary sulphur by being quite amorphous under the microscope, while the latter is crystalline.

Hydrosulphuric Acid (sulphuretted hydrogen) is known,—

1. By being a colorless gas with a disgusting odor of rotten eggs, and inflammable, burning in the air to produce sulphurous acid.
2. By turning a piece of paper black, which has been moistened with solution of plumbic acetate and held over the mouth of the tube or jet from which it issues.

Normal Sulphides are divisible into five classes:—

1. Soluble in water, including the sulphides of K, Na, NH_4, Ca, Sr, Ba and Mg.
2. Insoluble in water, but readily soluble in dilute hydrochloric acid, including those of Fe, Mn, Zn.
3. Insoluble in dilute, but soluble in strong boiling hydrochloric acid, including the sulphides of Ni, Co, Sb and Sn (PbS is also slightly affected, but separates on cooling, as chloride).
4. Insoluble in hydrochloric acid, but attacked by strong heated nitric acid, being converted wholly or partially into sulphates. These include the sulphides of Pb, Ag, Bi, Cu (arsenious sulphide is slowly affected).
5. Not dissolved by any single acid, but converted into a soluble sulphate by the action of nitro-hydrochloric acid, or hydrochloric acid and potassium chlorate; including those of Hg, As, Au, and Pt.

Sulphides soluble in water or in hydrochloric acid are recognised (practise on solution of Na_2S),—

1. By giving off sulphuretted hydrogen when heated with that acid, which gives the smell and reactions already noted.
2. Soluble sulphides precipitate solutions of lead and cadmium, black and yellow respectively.
3. Soluble sulphides give a purple color with sodium nitroprusside only after the addition of a soluble hydrate.

Sulphides insoluble in hydrochloric acid are best detected (practise on vermilion),—

1. By heating with strong nitric or nitro-hydrochloric acid, diluting the solution, and testing for a *sulphate* with barium chloride (see 24).
2. By fusion with $KNaCO_3$ and KNO_3, digesting the residue in water, filtering and testing the solution for a *sulphate* formed by the oxidising action of the potassium nitrate.
3. Mix a little with sodium carbonate and borax, and heat on charcoal before the blowpipe. Remove the mass thus obtained, place it on a clean silver coin, and moisten with a drop of distilled water; when, owing to the formation during ignition of sodium sulphide —Na_2S—a black stain of argentic sulphide—Ag_2S—will be produced.

Polysulphides as commonly met with are those of the alkalies, and are soluble in water. They are known (practise on sulphuretted potash),—

1. By the deep yellow or orange color of their solutions.
2. By evolving sulphuretted hydrogen *accompanied by a deposit of sulphur* when treated with hydrochloric or dilute sulphuric acids.

The polysulphides which are insoluble in hydrochloric acid, such as iron pyrites, copper pyrites, etc., are best proved by fusion with potassium nitrate and carbonate and conversion into sulphate. They may, however, be recognised by heating with hydrochloric acid and zinc, when the excess of sulphur will pass off as H_2S, leaving the normal sulphide.

20. DETECTION OF A SOLUBLE SULPHIDE IN PRESENCE OF A SULPHITE AND A SULPHATE.

(Practise on mixed solutions of Na_2S, Na_2SO_3, and Na_2SO_4.)

Pour the solution on a little cadmium carbonate—$CdCO_3$—filter, and treat the insoluble matter with acetic acid to remove any unacted-upon cadmium carbonate. If a sulphide have been present, a yellow residue of cadmium sulphide will remain insoluble in the acetic acid, while cadmium sulphite and sulphate will be found in the first filtrate, if these two radicals were present (see Separation of Sulphites and Sulphates).

21. THIOSULPHATES (Hyposulphites).

(Practise on solution of commercial hyposulphite of soda.)

These salts, commonly known as hyposulphites, are usually soluble in water, and exhibit the following characters:—

1. With either dilute or strong HCl and H_2SO_4, they yield SO_2 gas and a *yellow* deposit of S (*distinction from sulphides, polysulphides, and sulphites*).
2. **$AgNO_3$** gives no precipitate at first, owing to excess of a hyposulphite dissolving argentic hyposulphite—$Ag_2S_2O_3$—but on continuing the addition, this $Ag_2S_2O_3$ is precipitated of a *white* color. The salt splits up spontaneously, becoming yellow, brown, and lastly black, and being changed completely into argentic sulphide—Ag_2S. The same decomposition of the precipitate occurs on substituting $HgNO_3$ or $Pb(NO_3)_2$ for $AgNO_3$; and in all three cases heat accelerates the action, and H_2SO_4 is the by-product.
3. **Fe_2Cl_6, Na_2OCl_2**, and **Cl** water convert hyposulphites into sulphates, even without applying heat.

$$Na_2S_2O_3 + H_2O + 4Na_2OCl_2 = 2NaHSO_4 + 8NaCl.$$

The first produces a reddish-violet color, and this gradually disappears as $FeCl_2$ is formed. (*This color is not produced by sulphites.*)

22. SEPARATION OF THIOSULPHATES FROM SULPHIDES.

(Practise on solution of commercial hyposulphite of soda to which a drop of NH_4HS has been added.)

Having obtained a good preliminary idea by heating with H_2SO_4, add to a portion of the original solution—$ZnSO_4$—in excess, and filter.

(*a*) Precipitate white, and soluble in HCl, with smell of H_2S
= **Sulphides.**

(*b*) A portion of filtrate heated with H_2SO_4 deposits S and smells of SO_2; and another portion added to a drop or two of ammonio-cupric sulphate instantly causes decolorisation.
= **Hyposulphites.**

23. SULPHUROUS ACID and SULPHITES.

Sulphurous acid in solution is recognised,—

1. By its pungent odor of burning sulphur.
2. By adding barium chloride in excess, filtering out any precipitate

of barium sulphate which may form (owing to the fact that all samples of the ordinary acid contain sulphuric acid), and then adding chlorine water and getting another copious white precipitate of barium sulphate, owing to the conversion of the sulphurous into sulphuric acid by the oxidising action of the chlorine water, thus:—

$$H_2SO_3 + BaCl_2 + Cl_2 + H_2O = BaSO_4 + 4HCl.$$

3. Treated with zinc and hydrochloric acid, it evolves sulphuretted hydrogen, thus:—

$$3Zn + 6HCl + H_2SO_3 = 3ZnCl_2 + H_2S + 3H_2O.$$

4 When a solution of iodide is dropped into the liquid, its color is discharged, owing to its conversion into hydriodic acid by the hydrogen of the water, the oxygen of which passes at the same time to the sulphurous acid, forming sulphuric acid.

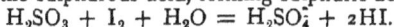

$$H_2SO_3 + I_2 + H_2O = H_2SO_4 + 2HI.$$

Sulphites are known by the following characteristic (practise on solution of sodium sulphite):—

1. All except the alkaline sulphites are sparingly soluble in water.
2. When heated with sulphuric acid they evolve sulphurous anhydride, *without deposit of sulphur.*
3. Acted on with zinc and hydrochloric acid, they evolve sulphuretted hydrogen, which blackens a piece of paper moistened with plumbic acetate and held over the mouth of the test-tube.
4. A salt of silver, mercury, or lead, produces a precipitate which on heating turns dark, owing to the formation of a sulphide and free sulphuric acid.
5. By boiling with barium chloride and chlorine water or nitric acid, barium sulphate is produced, and precipitates.
6. The acid gas—SO_2—combines directly with peroxides to form gaseous sulphates. For instance:—

$$PbO_2 + SO_2 = PbSO_4.$$

This reaction is utilised in gas analysis, to separate SO_2 from a mixture.

7. $K_2Cr_2O_7$ and HCl give a beautiful green coloration of *chromic sulphate* or *chloride.* This test is very delicate, but by itself is not conclusive.

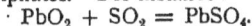

24. SULPHURIC ACID and SULPHATES.

The acid is detected,—

1. By its appearance. A heavy, oily, odorless, and nearly colorless liquid, powerfully acid and corrosive.
2. By its charring effect. This is made evident when the strong acid is dropped upon white paper, wood, etc., or when the dilute acid is evaporated in a basin containing a little white sugar. The carbonisation is due to the power the acid has of abstracting the elements of water from organic bodies.
3. By liberating an explosive gas from $KClO_3$ when heated with it. The test is not available with dilute acid.

Sulphates are soluble in water, with the exception of *basic sulphates* (soluble in acids) and $BaSO_4$, $SrSO_4$, $CaSO_4$, and $PbSO_4$. (Ag_2SO_4 is only slightly soluble.) When those insoluble in dilute acids are required to be analysed, they are decomposed either by boiling with potassium or sodium hydrates or by ignition with $KNaCO_3$, the latter being preferable. The sulphate radical, being brought into combination with K or Na to form a soluble sulphate, is to be tested for in the filtrate after boiling with water. Sulphates are recognised by the following characters (practise on solution of magnesium sulphate):—

1. **$BaCl_2$** or $Ba(NO_3)_2$ produces a white precipitate of barium sulphate —$BaSO_4$—insoluble in boiling water and also (after washing to remove excess of $BaCl_2$) in boiling nitric acid. This washing is necessary, as the addition of a strong acid to a solution of barium chloride often causes the reagent to crystallise out, and this is then mistaken by the student for a true precipitate of sulphate.
2. The addition of a soluble salt of lead or strontium also causes the formation of insoluble sulphates; but these reactions are never in practice used, the barium chloride being at once the most delicate and serviceable reagent.
3. Heated with a little Na_2CO_3 on charcoal in the inner blowpipe flame, sulphates are reduced to sulphides; and the residue placed on a clean silver coin and moistened with water, leaves a black stain.

25. SEPARATION of SULPHIDES, SULPHITES, and SULPHATES.

(Practise on mixed solutions of sodium sulphite and sulphate, to which a drop of NH_4HS has been added.)

Pour the solution on an excess of cadmium carbonate, digest at a gentle heat, filter, and examine the precipitate for a sulphide as already directed (19). The filtrate, which may contain the sulphite and sulphate, is precipitated by barium chloride, the insoluble precipitate filtered out and boiled with a little hydrochloric acid, which will dissolve the barium sulphite with evolution of sulphurous anhydride—SO_2—and leave the barium sulphate insoluble.

26. CARBON, CARBONIC ACID, and CARBONATES.

Carbon is known:—

1. By its black color and the production of a gas when burnt which is odorless, so heavy that it can be poured from one vessel to another, and causes a white precipitate when passed into solution of calcium hydrate.
2. By its capability of removing many vegetable coloring matters from their solutions.

Carbonic Acid is not known in the free state, splitting up into **carbonic anhydride**—CO_2—and water. CO_2 is recognised,—

1. By being odorless and giving white insoluble $CaCO_3$ (or $BaCO_3$) when passed into a solution of $Ca(HO)_2$ or $Ba(HO)_2$.
2. By turning blue litmus purple or wine-red, the original tint being restored by heat, the CO_2 escaping.

Carbonates are mostly insoluble in water, the alkaline carbonates alone dissolving. $CaCO_3$, $SrCO_3$, $BaCO_3$ and $MgCO_3$ (also $MnCO_3$ and $FeCO_3$), dissolve in water containing CO_2 (especially under pressure), forming bicarbonates or hydrogen carbonates, from which CO_2 passes off on boiling. All carbonates give off CO_2 on ignition, except K_2CO_3 and Na_2CO_3. A white heat is needed to decompose $BaCO_3$ and $SrCO_3$. Most carbonates on heating to redness leave the oxide. Their recognition depends upon (practise upon calcium carbonate):—

1. Effervescing with a solution of almost any acid (H_2S and HCy excepted), organic or inorganic, and giving off an odorless gas—CO_2.

2. When the gas given off is poured or passed into a solution of calcium hydrate, a white precipitate of $CaCO_3$ falls, which dissolves on continuing to add CO_2. When the gas given off has the odor of H_2S or SO_2, either of these may be removed by passing through K_2CrO_4 and HCl, which is rendered green, and the unacted-upon CO_2 is allowed to pass into calcium hydrate solution as before.

27. BORIC ACID and BORATES.

Boric (*or* **Boracic**) **Acid—H_3BO_3**—is distinguished as under:—

1. It is a white crystalline solid, giving off water on being heated, and leaving the *anhydride*—B_2O_3.
2. When warmed with alcohol, a *green flame* is produced on applying a light to the latter.
3. When dissolved in hot water, and a piece of turmeric paper dipped in the solution, the yellow color is unaffected; but upon drying the paper it becomes brownish-red.

All **borates** dissolve in dilute acids, but few in water, and when decomposed by hot acids, let fall crystalline boric acid on cooling, which answers to the above characters.

The presence of soluble **borates** is detected by the following tests (practise upon borax):—

1. They give, on heating with **calcium chloride**, rendered slightly alkaline with ammonium hydrate, a white precipitate of calcium borate, soluble in acetic acid, and so distinguished from oxalate.
2. On rendering the solution *just acid* with hydrochloric acid, and then dipping in a piece of turmeric paper, and drying the same after immersion at a *gentle* heat, it will be turned red.
3. Besides these two tests, which are in themselves, taken together, quite conclusive, borates give a precipitate with argentic nitrate soluble in nitric acid.
4. When a little of the solid borate is moistened, first with a drop of sulphuric acid and then with about a drachm of spirit of wine, the alcoholic solution of boric acid so obtained burns with a bright green flame.

28. SILICIC ACID and SILICATES.

The acid H_4SiO_4 is scarcely ever met with, and we have practically to deal with the anhydride—SiO_2—which is totally insoluble in water and dilute acids, the acid dissolving slightly in both. SiO_2 is characterised,—

1. By its infusibility when heated.
2. By its insolubility in water and all acids except HF.
3. By forming, when heated with H_2SO_4 and CaF_2 in a leaden vessel, *gaseous silicic fluoride*—SiF_4—which deposits the acid—H_4SiO_4—and forms *hydro-fluosilicic* acid—H_2SiF_6—in contact with moisture.

Silicates are not as a rule soluble in water, K_4SiO_4 and Na_4SiO_4 being the only ones thus affected, especially when much KHO or NaHO is present. Many of them do not dissolve in strong acids (a few are decomposed by hot H_2SO_4, but by no other acid), but all are split up by the action of gaseous hydrofluoric acid or a mixture of CaF_2 and H_2SO_4.

1. On adding HCl to a soluble silicate, H_4SiO_4 falls as a gelatinous

—scarcely visible—precipitate, slightly soluble in water. On evaporating to dryness and heating to 280° or 300° F., the addition to the residue of a little HCl and water leaves the SiO_2 as a white gritty powder.

2. NH_4Cl precipitates H_4SiO_4 from a soluble silicate.
3. Silicic anhydride is separated from all acidulous and basylous radicals by fusing the finely powdered insoluble silicate with Na_2CO_3 (or *fusion mixture*), in a platinum crucible; boiling with water, filtering, evaporating nearly to dryness, adding dilute HCl until the whole is acid, re-evaporating, moistening with water, and when dry heating to 280° or 300° F. On adding a little HCl, SiO_2 alone remains insoluble.

29. SEPARATION OF SILICIC ANHYDRIDE (SILICA) FROM ALL OTHER ACIDS.

(Practise upon powdered glass.)

Fuse the substance with a large excess of $KNaCO_3$ in a platinum crucible, and when all action has ceased, cool, and boil the residue with water. The silica passes into solution with the other acid radicals, and the metals are left as oxides. Acidulate the solution with hydrochloric acid, evaporate to dryness, and heat the residue to 280° Fah., and maintain the heat for some time. Drench the residue with strong hydrochloric acid, then add water, and boil, when the silica will alone remain insoluble.

30. HYDROFLUOSILICIC ACID (H_2SiF_6).

This acid is only known in solution.

1. It is very acid, and dissolves metals with the evolution of hydrogen, forming silico-fluorides which decompose by heat, leaving fluorides, and giving off silicon fluoride—SiF_4.
2. It gives off hydrofluoric acid when evaporated, and should not, therefore, be heated in glass vessels, as they would be etched.
3. The majority of silico-fluorides are soluble, the exceptions being K_2SiF_6, $BaSiF_6$, and Na_2SiF_6, which are insoluble, especially in presence of a little alcohol.
4. It does not precipitate strontium salts, even from strong solutions, but throws down $BaSiF_6$ on adding $BaCl_2$ and alcohol, as a *white translucent crystalline precipitate*.
5. Potassium salts throw down gelatinous K_2SiF_6.

31. NITROUS ACID AND NITRITES.

Nitrous Acid (so called commercially) is nitric acid containing nitrous anhydride. It is yellowish in color, and evolves reddish fumes.

Nitrites are all soluble in water, the least so being argentic nitrite. They are known as follows (practise upon potassium nitrate which has been heated to dull redness):—

1. They give red fumes when treated with strong sulphuric acid.
2. They give an instantaneous blue color with potassium iodide and starch paste on the addition of a few drops of dilute sulphuric acid. The sulphuric acid liberates hydriodic acid from the iodide, and nitrous acid from the nitrite; the hydriodic acid is decomposed by the nitrous acid into iodine, water, and nitric oxide:—

$$2HNO_2 + 2HI = I_2 + 2H_2O + 2NO.$$

[*Nitrates*, it must be remembered, would give frequently a similar reaction after standing, through the possible reduction of some portion of their nitric acid to nitrous acid ; so that unless the reaction appears instantly, and is confirmed by others, it is not safe to rely upon it as a test.]

3. They give a dark brown color with ferrous sulphate *without the previous addition of sulphuric acid, as required by nitrates.*
4. Potassium dichromate in solution is converted into a green liquid by the addition of a nitrite and an acid. These two latter substances also reduce an auric solution, forming a precipitate of the metal, possessing a dark color.

32. DETECTION OF A NITRITE IN THE PRESENCE OF A NITRATE.

(Practise upon potassium nitrate which has been slightly heated so as to partially decompose it.)

Add a little potassium iodide and starch paste, then introduce a small pinch of powdered metallic zinc, and lastly acidulate with acetic acid, when, if a nitrite be present, a blue color will be produced, due to the liberation of iodine. This test is often a very necessary one when dealing with drinking water, the presence in which of nitrites, derived from the oxidation of comparatively recent organic contamination, is a dangerous indication.

33. NITRIC ACID and NITRATES.

Nitric Acid is strongly acid and corrosive, fumes in the air, and readily dissolves most metals. It may be at once recognised by the following characters :—

1. When poured on a piece of copper foil, and a piece of white paper held behind the test-tube, it is observed to be filled with orange-red fumes of nitric peroxide—N_2O_4.
2. When dropped on a piece of quill in a basin, or if the solution be weak when evaporated in contact with the quill, the latter is stained yellow. This stain is intensified to orange on adding an alkali, and is not discharged by warming, both of which decolorise the corresponding stains produced by iodine and bromine.
3. Dropped on a few crystals of brucine, a deep red color is produced.

Nitrates are characterised by the following properties (practise upon solution of potassium nitrate) :—

1. All nitrates are soluble in water, especially when slightly acidulated with nitric acid. The nitrates of the alkalies are only decomposed by a very high temperature, but most of the nitrates of the heavy metals, such as copper, mercury, and lead, when heated are really decomposed, leaving a residue of oxide. Argentic nitrate, however, when heated leaves metallic silver.
2. When heated with sulphuric acid they evolve pungent fumes of nitric acid.
3. When heated with sulphuric acid and a piece of copper wire, they produce red fumes of nitric peroxide (caused by the union of the nitric oxide evolved with the oxygen of the air).
4. When mixed with a solution of ferrous sulphate in the presence of sulphuric acid, a black coloration is produced, which is due to the production of nitric oxide, and its absorption by the ferrous

salt. On heating, the color disappears, and the *ferrous* is changed to the *ferric* sulphate.

Note.—There are two ways of applying this test :—

(*a*) Place a drop or two of the suspected solution on a white porcelain slab or crucible lid, and having added a drop of strong sulphuric acid, put a small and clean crystal of the ferrous sulphate in the liquid, when a black ring will gradually form round the crystal.

(*b*) Place the suspected solution in a tube, and having added some strong solution of ferrous sulphate, cautiously pour some strong sulphuric acid down the side of the tube, so that it sinks to the bottom by reason of its great gravity without mixing with the fluid. If nitric acid be present, a dark line will be formed at the junction of the two liquids.

5. Treated with sulphuric acid, and a few drops of indigo sulphate added, the blue color of the latter is destroyed, being changed to yellow (not characteristic).

6. The most delicate test for nitrates is, however, phenyl-sulphuric acid. This reagent is prepared by dissolving one part of carbolic acid in four parts of strong sulphuric acid, and then diluting with two parts of water. A few drops of the solution to be tested are evaporated to dryness on a porcelain crucible lid over the water bath, and while still over the bath a drop of the reagent is added, when a reddish color is immediately produced, owing to the formation of nitro-phenol.

34. DETECTION OF FREE NITRIC ACID IN THE PRESENCE OF A NITRATE.

Digest with excess of barium carbonate ; filter, and add to the filtrate some dilute sulphuric acid, when, if the free acid were present, a precipitate of barium sulphate will be produced. This test is only good in the insured absence of sulphates and of any other acid capable of dissolving barium carbonate. It will also serve for detecting free hydrochloric and acetic acids in presence of their salts.

35. DETECTION OF A NITRATE IN THE PRESENCE OF AN IODIDE.

(Practise upon mixed solutions of potassium nitrate and potassium iodide.)

The fact that the addition of strong sulphuric acid liberates iodine renders the proof of a nitrate by the ordinary iron process doubtful in the presence of iodides and bromides. In this case proceed by one of the following methods :—

1. Boil with excess of potassium hydrate until any ammonium salts are decomposed, then add a fragment of zinc and again boil. Any nitrate present will be converted into ammonia, which may be recognised in the steam with moistened red litmus paper.

2. Warm with a little zinc amalgam and add a little acetic acid and starch paste, when the nitrate, being reduced to nitrite, will cause the liberation of iodine, and color the starch paste blue.

3. Boil with stannous chloride and a large excess of hydrochloric acid, which will produce ammonium chloride from a nitrate ; and by boiling the liquid with excess of potassium hydrate, the ammonia gas may be evolved. Of course the absence of ammonium salts must be first insured.

36. SEPARATION OF CHLORIDES, IODIDES, AND BROMIDES FROM NITRATES.

Digest with argentic sulphate, which will precipitate the halogens as silver salts and leave the nitrate in solution.

37. CYANOGEN, HYDROCYANIC ACID and CYANIDES.

Cyanogen is a colorless gas, which is recognised,—

1. By its odor of bitter almonds.
2. By its burning in the air with a peach-blossom-colored flame, producing carbonic anhydride and nitrogen.

Hydrocyanic Acid is volatile, soluble in water, and possesses a characteristic faint sickly odor of almonds. Its reddening action on litmus paper is very transient. Its tests are four in number, as follows:—

1. *The Silver Test.*—Argentic nitrate added to a solution of prussic acid gives a curdy white precipitate of argentic cyanide. The precipitate is soluble in ammonium hydrate and in strong boiling nitric acid, but not in dilute nitric acid; nor does it blacken on exposure to the light.
2. *Scheele's Iron Test.*—An excess of solution of potassium hydrate is mixed with the solution. To this a mixture of a ferrous and a ferric salt is added, and the whole acidulated with hydrochloric acid. If hydrocyanic acid be present, *Prussian blue* will be formed.

The explanation of the test is as follows (according to Gerhardt's view):—
(1) The hydrocyanic acid and the potassium hydrate form potassium cyanide.
(2) The addition of the ferrous salt produces ferrous cyanide.
(3) This reacting with the excess of alkali forms potassium ferrocyanide.
(4) On the addition of the ferric salt, it is at first precipitated by the excess of alkali, as ferric hydrate, which on acidulation dissolves to ferric chloride, forming ferric ferrocyanide (Prussian blue).

(i) $6HCy + 6KHO = 6KCy + 6H_2O.$
(ii) $6KCy + 3FeCl_2 = 3FeCy_2 + 6KCl.$
(iii) $F3eCy_2 + 4KHO = K_4FeCy_6 + 2Fe(HO)_2.$
(iv) $3K_4FeCy_6 + 2Fe_2Cl_6 = (Fe_2)_2(FeCy_6)_3 + 12\ KCl.$

Or the whole may be shown in one equation, thus, which is quite sufficient:—
$18KCy + 3FeCl_2 + 2Fe_2Cl_6 = (Fe_2)_2(FeCy_6)_3 + 18\ KCl.$

3. *The Sulphur Test.*—A few drops of *yellow* ammonium sulphydrate is added to a solution of hydrocyanic acid, and the whole evaporated to dryness at a very gentle heat, with the addition of a drop of ammonium hydrate. A residue is thus obtained which strikes a blood-red color with ferric chloride, not dischargeable by hydrochloric acid, but at once bleached by solution of mercuric chloride.

This color is due to the formation of ammonium sulphocyanide (which takes place when an alkaline sulphide, containing excess of sulphur, is brought into contact with cyanogen)—
$$2HCy + (NH_4)_2S + S_2 = 2NH_4CyS + H_2S,$$
and subsequent production of red ferric sulphocyanide.

4. *Schönbein's Test.*—It has been stated that a very delicate means of detecting HCy is based upon its action on filtering paper, soaked, first in a 3 per cent. alcoholic solution of guaiacum resin, and then in a 2 per cent. solution of cupric sulphate, and exposed to the air. The presence of HCy causes the production of a blue color. The paper may be either moistened with the suspected solution or exposed to its vapor.

Cyanides are known (practise upon solution of potassium cyanide),—

1. By effervescing and giving off the odor of hydrocyanic acid when heated with sulphuric acid.
2. By answering to all the tests for hydrocyanic acid above mentioned.

Note.—In using the silver test to a soluble cyanide, the reagent must be added in excess, as argentic cyanide is soluble in alkaline cyanides to form double cyanides of silver and the alkali used. Excess of argentic nitrate, however, decomposes these compounds, and forms insoluble argentic cyanide. The previous addition of a *slight* excess of dilute nitric acid ensures the immediate separation of the argentic cyanide, by preventing the reaction just referred to.

3. Insoluble cyanides yield cyanogen when heated *per se* in a small dry test-tube, the open end of which has been drawn out into a jet after the introduction of the cyanide. The application of a light to the jet causes the characteristic flame of cyanogen.

38. SEPARATION OF CYANIDES FROM CHLORIDES.

(Practise on a mixed solution of KCN and KCl.)

Acidulate slightly with nitric acid, add *excess* of argentic nitrate, wash the precipitate thoroughly with boiling water by decantation, allow it to settle completely, pour off all the water, and boil with strong nitric acid, when the argentic cyanide is decomposed, leaving the chloride insoluble. The solution in nitric acid is diluted and hydrochloric acid added, when any dissolved silver is detected by precipitation, as chloride, thus indicating the presence of argentic cyanide in the original mixture.

39. CYANIC ACID and CYANATES, CYANURIC ACID, and FU L INIC ACID.

Cyanic acid is characterised,—

1. By being a colorless liquid, having a strong pungent odor, greatly resembling acetic acid, or sulphurous acid when in small quantity, and forming ammonium bicarbonate on adding water.
2. By changing into a white solid *isomer* on keeping, heat being evolved, but no decomposition occurring.

Cyanates are known,—

1. By giving, when moistened, a bicarbonate. (The potassium salt—KCyO—for instance, forms potassium bicarbonate—$KHCO_3$.)
2. By producing urea when evaporated with an ammonium salt.

Cyanuric acid is a polymeric modification of cyanic acid, which is recognised,—

1. By being a crystalline solid, yielding cyanic acid on applying heat.
2. By not being decomposed by strong hot HNO_3 or H_2SO_4.

Fulminic acid (intermediate between the two above acids) differs from both by the fearful explosibility of its salts.

40. THIOCYANATES (Sulphocyanates).

Sulphocyanates are recognised,—

1. By being usually colorless and soluble, and evolving hydrocyanic acid and depositing sulphur on heating with sulphuric acid.
2. By producing with a ferric salt, a blood-red solution of *ferric sulphocyanate*—$Fe_2(CyS)_6$—the color of which is notdestroyed by HCl, but disappears on adding mercuric chloride—$HgCl_2$.

41. FERROCYANIDES.

(Practise upon solution of potassium ferrocyanide.)

Ferrocyanides are mostly insoluble in water, except those of the metals of the first and second groups. They are characterised,—

1. By giving off hydrocyanic acid and forming a deposit on heating with sulphuric acid.
2. By giving with a **ferrous salt** a white precipitate of ferrous potassium ferrocyanide—$K_2Fe(FeCy_6)$—changing quickly to blue.

3. By yielding with a **ferric salt** a dark blue precipitate of *ferric ferrocyanide*—$(Fe_2)_2(FeCy_6)_3$. This precipitate and (1) are both insoluble in HCl, but KHO decomposes them, producing potassium ferrocyanide. When it is boiled with HgO and water, mercuric cyanide—$HgCy_2$—and Fe_2O_3 remain.
4. **Cupric salts** produce a reddish-brown precipitate of cupric ferrocyanide—Cu_2FeCy_6—insoluble in acids, dissolved by NH_4HO, but left unaltered on evaporating off the ammonia.
5. By precipitating a white ferrocyanide from a solution of a lead salt.
6. By yielding a white mercuric ferrocyanide in a mercuric solution.
7. By giving no effect with magnesium salts but a white gelatinous precipitate, soluble in NH_4HO on the addition of a solution containing a zinc salt.
8. By producing with argentic nitrate—$AgNO_3$—white gelatinous silver ferrocyanide, dissolved by NH_4HO.

None of these precipitates can be produced in alkaline solutions; and they form best in slightly acid solutions.

42. FERRICYANIDES.

(Practise upon solution of potassium ferricyanide.)

Most ferricyanides are insoluble, those of the alkalies and of the barium group being exceptions. They are recognised,—

1. By yielding an odor of hydrocyanic acid, and a deposit on heating with sulphuric acid.
2. By producing with a **ferrous** salt dark-tinted *Turnbull's blue*—$Fe_3Fe_2Cy_{12}$—insoluble in acids, but forming $K_6Fe_2Cy_{12}$ when boiled with KHO.
3. By producing a brownish coloration when added to a **ferric** salt in solution, from which H_2SO_3, *Stannous Chloride*—$SnCl_2$—and other reducing agents throw down *Turnbull's blue* or *Prussian blue* (*distinction between* $H_6Fe_2Cy_{12}$ *and* H_4FeCy_6).
4. By giving no precipitate in a lead solution (*another distinction of a ferricyanide from a ferrocyanide*).
5. By throwing down *mercurous ferricyanide* of a brownish-red color from a mercurous solution.
6. By yielding with argentic nitrate solution an *orange* precipitate of argentic ferricyanide.
7. Mercuric salts, giving no precipitate.
8. Stannous salts, a white precipitate, soluble in HCl.
9. Stannic salts, no visible alteration.

43. SEPARATION OF FERRO- FROM FERRI-CYANIDES.

(Practise upon mixed solutions of potassium ferro- and ferricyanides.)

Acidulate with hydrochloric acid, add excess of ferric chloride, warm gently; the ferrocyanide will be precipitated. Pour off some of the brownish liquid and heat with a little zinc amalgam, when a blue precipitate will form, owing to the reduction of the ferri- to ferro-cyanide.

44. DETECTION OF CYANIDES IN THE PRESENCE OF FERRO- AND FERRI-CYANIDES.

(Practise upon mixed solutions of potassium cyanide and ferrocyanide.)

Acidulate slightly with nitric acid and add an excess of a mixture of ferrous and ferric sulphates; warm gently, and allow the precipitate to subside. Pour off a little of the supernatant liquid, add to it excess of potassium hydrate, and then acidulate with hydrochloric acid, when the production of another blue precipitate will prove the cyanide.

45. HYPOPHOSPHITES.

(Practise upon solution of calcium hypophosphite.)

The silver salt alone is insoluble in water, and few are insoluble in alcohol. The following reactions serve for their detection :—

1. When heated in a solid state, they take fire, evolving phosphuretted hydrogen, and leaving a residue of pyrophosphate.

   Note.—This must be done on porcelain, as they destroy platinum foil.
2. With argentic nitrate they give a white precipitate, which turns brown owing to its reduction to metallic silver.
3. With mercuric chloride they yield, when slightly acidulated with HCl, a precipitate of calomel, which, on heating, turns dark, owing to a reduction to the metallic state.
4. After removal of the base, the free hypophosphorous acid, when boiled with solution of cupric sulphate, will give a deposit of metallic copper.
5. Treated with ammonium molybdate they give a fine *blue* precipitate. As afterwards mentioned, *phosphates* give a *yellow*, and consequently when the solution contains both classes of salts the precipitate is *green*. This forms an excellent and rapid method of checking any commercial sample of hypophosphites.

They are distinguished from phosphites by not giving precipitates with neutral barium, or calcium, chlorides, or with plumbic acetate. In performing the 4th reaction, the base, if calcium, is removed by oxalic acid, if barium, by sulphuric acid, and if a heavy metal, by sulphuretted hydrogen.

46. PHOSPHOROUS ACID and PHOSPHITES

are distinguished as follows :—

1. Heated on platinum foil, they burn. They are powerful reducing agents.
2. The only phosphites soluble in water are those of K, Na, and NH_4, but acetic acid dissolves all, except plumbic phosphite.
3. With zinc and sulphuric acid (nascent hydrogen) they yield phosphuretted hydrogen, burning with an emerald-green color, and throwing down Ag_3P, as well as Ag from $AgNO_3$ in solution.
4. They give precipitates with neutral barium, and calcium chlorides, and also with plumbic acetate, which hypophosphites do not.
5. Heated with mercuric chloride or argentic nitrate, they yield a precipitate of metallic mercury or silver.

$$2HgCl_2+2H_3PO_3+2H_2O=2H_3PO_4+Hg_2+4HCl.$$

47. META- AND PYRO-PHOSPHORIC ACIDS AND THEIR SALTS.

Metaphosphoric Acid is a glassy solid, not volatile by heat. It is freely soluble in cold water, and is converted by boiling into orthophosphoric acid. It is known by giving a white precipitate with argent-ammonium nitrate, and by its power of coagulating albumen.

Metaphosphates are known by,—

1. Giving no precipitate with ammonium chloride, ammonium hydrate, and magnesium sulphate, added successively.
2. By giving a white precipitate of argentic metaphosphate—$AgPO_3$—with argentic nitrate only in neutral solutions, and soluble both in nitric acid and ammonium hydrate.

Pyrophosphoric Acid is also soluble in water and convertible by boiling into orthophosphoric acid. It gives a white precipitate with argent-ammonium nitrate, but does not coagulate albumen. Pyrophosphates are insoluble in water, except those of the alkalies. Their tests are not very well defined, but they give,—

1. A white precipitate of argentic pyrophosphate—$Ag_4P_2O_7$—with argentic nitrate in a neutral solution only and soluble both in nitric acid and ammonium hydrate.
2. $(NH_4)_2MoO_4$ does not produce an immediate precipitate.

48. ORTHOPHOSPHORIC ACID and ORTHOPHOSPHATES.

Orthophosphoric acid is a liquid with a strongly acid reaction, converted by heat first into pyro- and finally into meta-phosphoric acid, which remains as a glassy residue. It is,—

1. Not volatile by a red heat.
2. It gives a yellow precipitate of argentic phosphate—Ag_3PO_4—when treated with argent-ammonium nitrate, soluble both in nitric acid and ammonium hydrate.

Phosphates are as a rule insoluble in water, except the alkaline ones. They are readily soluble in dilute acids, and entirely reprecipitated on neutralising by an alkali or alkaline carbonate. Calcium, strontium, and barium phosphates are only partly soluble in dilute sulphuric acid, being converted into a soluble phosphate and an insoluble sulphate of the metal. If the insoluble phosphate be filtered out, the addition of an alkali causes only a slight precipitate of a dimetallic phosphate, and a phosphate of the alkali used is left in solution; but it is only after the use of sulphuric acid that any phosphate thus remains dissolved.

Phosphates are detected as follows (practise on solution of sodium phosphate):—

1. With **barium** or **calcium chloride** white precipitates are produced, soluble in acetic acid (distinction from oxalates) and all stronger acids.
2. With **argentic nitrate** a lemon-yellow precipitate of argentic phosphate forms, soluble both in nitric acid and ammonium hydrate.
3. With **ferric chloride** in the presence of **ammonium acetate** a white precipitate of **ferric phosphate** appears, insoluble in acetic acid.
4. With **magnesia mixture** phosphates yield a white crystalline precipitate, forming slowly in dilute solutions, consisting of **ammonium-magnesium phosphate**—$Mg(NH_4)PO_4+6H_2O$—soluble in acetic and all acids.
5. With solution of **ammonium molybdate** in **nitric acid** a yellow precipitate is produced, insoluble in nitric acid, but soluble in **ammonium hydrate.**
6. With **uranic nitrate** phosphates yield a yellow precipitate of *uranic phosphate*, also insoluble in acetic acid.
7. With **mercurous** and **bismuthous nitrates** white precipitates are formed, the former soluble and the latter insoluble in nitric acid.
8. With any **soluble salt of lead** a white precipitate of plumbic phosphate is produced, soluble in nitric acid, but insoluble in acetic acid or ammonium hydrate.

It is important to note that all these reactions may be caused by **arsenic acid**, except that with argentic nitrate, argentic arseniate differing from argentic phosphate in being brick-red instead of yellow.

49. DETECTION OF A PHOSPHATE IN THE PRESENCE OF CALCIUM, BARIUM, STRONTIUM, MANGANESE, AND MAGNESIUM.

(Practise on ordinary "phosphate of lime.")

Dissolve in water by the aid of the smallest quantity of nitric acid, then add excess of ammonium acetate, which will remove the excess of nitric acid without rendering the solution alkaline. In this solution the phosphate may be proved by adding a drop or two of ferric chloride and warming, when a white precipitate of ferric phosphate—$Fe_2(PO_4)_2$—will form, insoluble in the acetic acid liberated.

50. DETECTION OF A PHOSPHATE IN THE PRESENCE OF IRON.

(Practise on ordinary "phosphate of iron.")

Dissolve in the smallest possible quantity of hydrochloric acid, add some citric acid, and then excess of ammonium hydrate. By this means an alkaline liquid is obtained, owing to the power of the organic acid to prevent the precipitation of the metal by the ammonium hydrate; and in this liquid, when cold, *magnesia mixture* (ammonio-sulphate of magnesia) causes the precipitation of white crystalline magnesium-ammonium phosphate.

51. ARSENIOUS ACID and ARSENITES.

Arsenious Acid—H_3AsO_3—is not known in the free state; but its anhydride —As_2O_3—is common. The latter has the following qualities :—

1. Dropped upon red-hot charcoal or coal, or heated in a dry tube with black flux, or a mixture of dry sodium carbonate and potassium cyanide, the metalloid **As_4** is set free, and volatilises with a garlic odor, producing a steel-grey mirror on the sides of the tube.
2. Dissolved in water only, and argent-ammonium nitrate added, a canary-yellow precipitate of argentic arsenite—Ag_3AsO_3—is produced, soluble in excess of either NH_4HO or HNO_3.
3. A pure aqueous solution, mixed with cupr-ammonium sulphate, gives a bright-green cupric arsenite—*Scheele's green*—also soluble in NH_4HO or in HNO_3.
4. Any solution yields all the tests for **arsenic** (see page 15).

Arsenites behave peculiarly in many respects. Ammonium arsenite leaves arsenious acid on evaporating a solution, while potassium and sodium arsenites possess a degree of alkalinity which no excess of arsenious acid will disturb. Ba, Sr, and Ca form soluble hydrogen salts. All other arsenites are insoluble.

Neutral solutions of **arsenites** are possessed of the undermentioned distinctive peculiarities :—

1. $CuSO_4$ throws down greenish cupric arsenite.
2. $AgNO_3$ is transformed into yellow insoluble argentic arsenite.
3. H_2S *in the presence of hydrochloric acid*, gives a yellow precipitate of arsenious sulphide.
4. The solution gives the usual tests for **arsenic** (see page 15).

52. ARSENIC ACID and ARSENIATES.

Arsenic Acid—H_3AsO_4—is known by the following characters :—

1. The crystals are deliquescent, white, and strongly acid. Heated, they leave a residue, which, on moistening with water, is also acid.

2. It is strongly corrosive and blisters the skin. It gives *brick-red* Ag_3AsO_4 on adding *argent-ammonium nitrate.*

Arseniates are characterised by behaving in every respect exactly like phosphates, except that they give a brick-red precipitate with argentic nitrate, instead of a yellow. Insoluble arseniates are best treated by boiling with NaHO, filtering, *exactly neutralising* the filtrate with dilute HNO_3, and then getting the brick-red precipitate with $AgNO_3$.

53. SEPARATION OF AN ARSENIATE FROM A PHOSPHATE.

This can only be done by acidulating with hydrochloric acid, and passing a slow stream of sulphuretted hydrogen through the solution for several hours, until the whole of the arsenic is removed.

54. MANGANATES.

Manganates are unstable compounds, and only the alkaline salts dissolve in water, forming green solutions.

1. Soluble manganates decompose spontaneously, depositing MnO_2, the green color changing to purple or reddish violet, owing to the formation of a permanganate.

$$3K_2MnO_4 + 2H_2O = 2KMnO_4 + 4KHO + MnO_2.$$

2. Dilute acids effect this change more rapidly, and the reaction is very delicate. The free hydrate is then replaced by a chloride, nitrate, or sulphate.
3. Strong, heated H_2SO_4 acts as represented in this equation :—

$$K_2MnO_4 + 2H_2SO_4 = K_2SO_4 + MnSO_4 + 2H_2O + O_2.$$

4. Strong HCl causes the evolution of Cl_2. The other actions are similar to those of permanganates, but less energetic.

55. PERMANGANATES.

(Practise on solution of potassium permanganate.)

Permanganates are known,—

1. By the violet color of their solutions, which is entirely bleached by oxalic acid or by heating with hydrochloric acid and dropping in rectified spirit.
2. By giving off oxygen on heating.
3. By giving off oxygen when heated with sulphuric acid, often with explosive violence.
4. By evolving chlorine when simply mixed with hydrochloric acid.

56. CHROMIC ACID and CHROMATES.

Chromic Acid—H_2CrO_4—not being definitely proved to exist, is represented by its *anhydride*—CrO_3. This is a dark red crystalline solid, and when mixed with an aqueous solution of *hydrogen* peroxide—H_2O_2—a *deep blue* liquid results, which is believed to contain *perchromic acid*—$HCrO_4$ or $H_2Cr_2O_8$. The liquid decomposes rapidly, unless ether be added, which lengthens its existence.

This test, for either CrO_3 or H_2O_2, is exceedingly delicate, the ethereal solution of perchromic acid separating from the water and thus concentrating the color into a small bulk of ether.

Chromates of the alkalies are soluble, while those of the other metals are chiefly insoluble, but have very brilliant colors. They are very poisonous, and are detected as follows (practise on solution of K_2CrO_4 and $K_2Cr_2O_7$ respectively):—

1. **Soluble chromates** give a yellow precipitate with plumbic acetate or barium chloride, soluble in nitric acid, insoluble in acetic acid. The lead salt is darkened in color by alkalies, and dissolves by free excess of hot KHO.
2. With argentic nitrate a dark red precipitate, also soluble in nitric acid, and in NH_4HO, but not in acetic acid—$HC_2H_3O_2$.
3. Boiled with hydrochloric acid and alcohol, or any reducing agent, (for instance, sulphurous acid—H_2SO_3), they turn green, owing to the production of chromic chloride (or sulphate).
4. Treated with sulphuretted hydrogen, in the presence of hydrochloric acid, they turn green, and a deposit of sulphur takes place:—
$$2K_2Cr_2O_7 + 16HCl + 6H_2S = 2Cr_2Cl_6 + 4KCl + 3S_2 + 14H_2O.$$
5. Chromates, when treated with an acid, turn orange; and dichromates, when treated with potassium hydrate or any alkali, turn yellow. In this way they are mutually distinguished.
6. Heated with strong H_2SO_4 they give off oxygen.
7. Treated with an excess of sulphuric acid, and shaken up with ozonised ether (solution of hydric peroxide in ether), they give a gorgeous blue with the most minute traces.

57. STANNIC ACID and STANNATES (Stannites ?).

This is an unimportant compound, and is thrown down by an alkaline hydrate from a stannic salt. It is sometimes stated to be endowed with the composition $Sn(HO)_4$, and at others, $SnO(HO)_2$ (H_2SnO_3).

Stannates are formed by the solution of the acid in an alkaline hydrate, and are detected in the examination for metals.

Stannites are said to be formed by the solution of stannous hydrates in an alkaline hydrate. They decompose on boiling with KHO, forming *stannates* and throwing down metallic tin.

58. ANTIMONIC ACID.

This is the white precipitate having the composition $HSbO_3$, formed on adding strong HCl to *potassium antimoniate*; and it is detected in the examination for metals.

59. FORMIC ACID and FORMATES.

Formic Acid—$HCHO_2$—is the "organic" acid which contains the highest percentage of oxygen, and approaches most nearly in composition to the suppositious carbonic acid—H_2CO_3. It is a tolerably stable liquid, boiling at the same temperature as water. Formates are all soluble in water, and behave as follows:—

1. Heated to redness they decompose without blackening.
2. Heated with H_2SO_4 they evolve CO, which, being free from CO_2, gives no effect when passed through *lime-water*, but burns with the usual pale blue flame. The reaction is:—
$$H_2SO_4, + HCHO_2 = H_2SO_4 + CO + H_2O.$$
3. Readily reduce argentic nitrate, when boiled, metallic silver separating.

60. DETECTION OF A FORMATE IN THE PRESENCE OF FIXED ORGANIC ACIDS WHICH REDUCE SILVER SALTS.

Distil with dilute sulphuric acid at the heat of a water bath, neutralise the distillate with sodium carbonate, add a *slight* excess of acetic acid, and boil with argentic nitrate, when a dark deposit of metallic silver will form.

61. ACETIC ACID and ACETATES.

(Practise on sodium acetate.)

This acid is characterised by its odor of vinegar. The strong acid chars when heated with strong H_2SO_4.

Acetates are without exception soluble in water ($Ag_2C_2H_3O_2$ and $HgC_2H_3O_2$ are sparingly dissolved). They are decomposed by a red heat, yielding *acetone* if the heat rise gently and the mass be not alkaline, and leaving a carbonate, oxide, or metal, according to the nature of the basylous radical. When heated with alkalies *marsh gas*—CH_4,—is evolved. The reaction is of this type:—

$$NaC_2H_3O_2 + NaHO = Na_2CO_3 + CH_4.$$

In the case of no hydrate or carbonate being present, the following is an example of the effect of heat on acetates:—

$$Ba(C_2H_3O_2)_2 = BaCO_3 + C_3H_6O \text{ (Acetone)}.$$

Acetates of easily reducible metals, such as copper, yield, when heated, a distillate of acetic acid, leaving a residue of the metal, or in some cases of oxide. The presence of **acetates** is analytically determined as follows:—

1. By evolving an odor of acetic acid when heated with sulphuric acid.
2. By a characteristic apple-like odor of "acetic ether"—$C_2H_5(C_2H_3O_2)$—which they evolve when heated with sulphuric acid and rectified spirit.
3. By the deep red color which they produce with neutral ferric chloride—ferric acetate, $Fe_2(C_2H_3O_2)_6$—dischargeable by both hydrochloric acid and mercuric chloride.

62. VALERIANIC ACID and VALERIANATES.

Valerianic Acid is a liquid, which is—

Volatile, malodorous, colorless, and oily. It reddens test-paper, and dissolves in most *menstrua*.

The general characters of **Valerianates** are:—

1. A more or less strong odor of valerian root when warmed or moistened.
2. They give, when heated with sulphuric acid, an odor of valerian and a distillate which, on the addition of solution of cupric acetate, forms, after the lapse of some time, an oily precipitate; gradually solidifying, by the absorption of water, into a greenish-blue crystalline solid.

63. SULPHOVINATES (Ethyl sulphates).

These salts, derived from ethyl hydrogen sulphate, behave as follows:—

1. Heated with strong sulphuric acid, they evolve a faint ethereal odor.
2. They give no precipitate in the cold with barium chloride; but on boiling, a white precipitate of barium sulphate falls, and a smell of alcohol is perceived. The addition of a little solution of barium hydrate after the chloride and before boiling facilitates the reaction; but in this case all metals precipitable by a fixed alkali must first, of course, be removed.
3. Heated to redness, they leave a sulphate of the metal.
4. Heated with sulphuric acid and an acetate, or with strong acetic acid, they evolve acetic ether with its characteristic apple odor.

64. STEARIC ACID and STEARATES.

This acid is usually so distinguished by its appearance and behaviour on being heated, that further tests are useless. The characters are :—

1. A white, odorless, fatty solid, melting by heat and soluble in absolute alcohol, the solution having an acid reaction.
2. Giving, when dissolved in KHO and the solution as nearly neutralised as possible, a white insoluble precipitate of plumbic stearate —$Pb(C_{18}H_{35}O_2)_2$—on the addition of plumbic acetate, which is insoluble in ether (distinction from plumbic oleate).

Stearates of the alkalies are alone soluble in water.

Any stearate heated with dilute HCl gives the free acid, which floats as an oily liquid, solidifying on cooling to a white mass. This test is applicable to the analysis of soap (*hard*, containing Na, or *soft*, in which K is present).

65. OLEIC ACID and OLEATES.

Oleic Acid is usually an oily liquid, but remains solid below 59° F. when crystallised from alcohol.

It does not dissolve in water, but is taken up by ether and by strong alcohol, the latter solution being acid in reaction.

Oleates of **K** and Na alone dissolve in water. Acid oleates are all liquid and soluble in cold absolute alcohol and ether.

1. They do not separate out from either of these solvents when a hot solution is cooled (*distinction from stearates and palmitates*).
2. *Plumbic oleate* is precipitable like plumbic stearate, but is separated and distinguished from it by dissolving in ether.

66. LACTIC ACID and LACTATES.

The pure strong acid resembles *glycerine* in appearance, liberates hydrogen on adding zinc, and on heating takes fire and burns away with a pale flame gradually becoming luminous. It dissolves in ether. It gives pure CO when heated with sulphuric acid. Boiled with solution of potassium permanganate it gives the odor of aldehyd.

Lactates are not very soluble in water. They—

1. Are insoluble in ether.
2. *Argentic lactate*—$AgC_3H_5O_3$—when boiled gives a dark precipitate, which on subsidence leaves a *blue liquid*.
3. Strong solution of an alkaline lactate, when boiled with $HgNO_3$, deposits *crimson* or *pink* mercurous lactate—$HgC_3H_5O_3$.

67. OXALIC ACID and OXALATES.

(Practise on oxalic acid and on "salts of sorrel.")

The acid is very common, and is recognised,—

1. By its colorless prismatic crystals, which are strongly acid, effloresce when exposed to dry air, and volatilise on heating with partial decomposition.
2. By the complete discharge it effects of the color of a solution of potassium permanganate acidulated with dilute H_2SO_4.

3. By producing free H_2SO_4 when added to solution of $CuSO_4$. (This is one of the very rare instances in which SO_4 is replaced by another acid radical and H_2SO_4 liberated.)
4. By giving the reactions of an oxalate.

Oxalates of the alkalies are soluble, the others insoluble, in water. Insoluble oxalates dissolve in hydrochloric but not in acetic acid. They are known by—

1. Not charring when heated, but only turning faintly grey; followed by a sudden glow of incandescence, which runs through the mass.
2. Not charring when heated with sulphuric acid, but yielding CO and CO_2 with effervescence.
3. Not effervescing with cold dilute sulphuric acid; but at once liberating CO_2 with effervescence on the addition of a pinch of manganese peroxide.
4. With calcium chloride or barium chloride in a neutral or alkaline solution, they give a white precipitate of calcium or barium oxalate, *insoluble in acetic acid, but soluble in hydrochloric acid.*

(For separation of oxalates from tartrates, etc., see No. 72.)

68. SUCCINIC ACID and SUCCINATES.

This acid is a white crystalline solid. It is known,—

1. By not charring with strong hot sulphuric acid.
2. By subliming in a tube open at both ends, in silky needles, *without giving off an irritating vapor* (distinction from benzoic acid).
3. By burning, when heated on platinum, with a blue smokeless flame.

Succinates are recognised as follows:—

1. With ferric chloride, a brownish-red precipitate of *ferric succinate*—$Fe_2(C_4H_4O_4)_3$—is formed.
2. With hydrochloric and sulphuric acids no precipitate is produced (*distinction from benzoates*). With *plumbic acetate*, a white precipitate of *plumbic succinate*, soluble in succinic acid, succinates, and plumbic acetate.
3. *Barium succinate* is soluble in hydrochloric acid, hence no effect results from the addition of succinic acid to barium chloride; but alcohol and ammonium hydrate give rise to a white precipitate (*another point of distinction from benzoates*).

69. MALIC ACID and MALATES.

Malic Acid—$H_2C_4H_4O_5$—is a colorless, crystalline, very deliquescent acid, freely soluble in water and alcohol. Acid **malates** are most stable. The characters are:—

1. Calcium chloride, added to a neutral solution of a malate, gives no precipitate. Alcohol, however, even if added in small quantity, throws down a white precipitate; and boiling aids the effect.
2. Strong H_2SO_4 gives no charring for some time (*a tartrate is carbonised in a few minutes*).
3. Amorphous *plumbic malate* fuses below 100° C. in water, but not in an air-bath.

70. TARTARIC ACID and TARTRATES.

(Practise upon the free acid and also upon "Rochelle salt.")

Tartaric Acid—$H_2C_4H_4O_6$— is a strong acid, soluble in water and spirit.

1. It forms usually oblique rhombic prismatic crystals, of an acid taste.
2. It is decomposed by heat, giving off the odor of *burnt sugar*, and leaving carbon. A similar effect is produced by warming with strong H_2SO_4, which blackens and carbonises it in a few minutes.
3. With potassium acetate—$KC_2H_3O_2$—it produces a white crystalline precipitate of *potassium hydrogen tartrate*—$KHC_4H_4O_6$—in either an aqueous or an alcoholic solution, soluble in much water, but not in spirit. Stirring or violent shaking promotes the formation of the salt.

The same compound is produced on adding any potassium salt, provided the liquid contain excess of free tartaric acid only.

Tartrates of the alkalies are mostly soluble; but the others are insoluble. The hydrogen tartrates of K and (NH_4) are nearly insoluble. Tartrates are recognised by the following characters :—

1. Heated to dull redness they char rapidly and give off a smell of burnt sugar. The black residue contains the metal as carbonate if it be K, Na, Li, Ba, Sr, or Ca; but the tartrates of other metals usually leave the oxides, or more rarely (as in the case of $Ag_2C_4H_4O_6$) the metal.
2. Heated with strong sulphuric acid, they blacken rapidly, and give first a smell of burnt sugar, and afterwards evolve SO_2.
3. Neutral solutions (free from more than a trace of ammonium salts) give, on adding **calcium chloride**, a white precipitate of *calcium tartrate*, which, when freed from other salts by washing, dissolves readily in cold solution of potassium hydrate, but is again precipitated on boiling. The precipitate is soluble in NH_4Cl, but not in NH_4HO.
4. Mix a tartrate with **sodium carbonate**, and filter the slightly alkaline solution, so that the only metal present shall be sodium. If this clear solution, after *slight* acidulation with acetic acid, be mixed with argentic nitrate, the whole, on heating nearly to boiling in a clean tube, deposits *a beautiful mirror of metallic silver* upon the test-tube employed.
5. A tartrate prevents the precipitation (more or less perfectly) of the salts of Pb, Bi, Cd, Cu, Pt, Fe_2, Mn, Ni, Co, Cr, and Zn, by an alkaline hydrate in excess, or a phosphate.

71. CITRIC ACID and CITRATES.

(Practise upon the free acid and upon potassium citrate.)

Citric Acid—$H_3C_6H_5O_7$—is soluble in water and alcohol, but insoluble in pure ether. It entirely burns away when heated to redness in the air; blackens slowly when heated with strong sulphuric acid; and when neutralised by ammonium hydrate, and the solution cooled, the solution gives no precipitate with calcium chloride until it has been boiled. Added to ferric, chromic, or aluminic salts in solution, it prevents their precipitation by ammonium hydrate.

Citrates exhibit the following characters :—

1. Heated alone, they char slowly, and evolve an odor of burnt sugar, but not so intense as that of a tartrate. At a dull red heat, the citrates of K, Na, Li, Ba, Sr, and Ca leave their carbonates; but those of most other metals leave the oxides. Argentic citrate leaves the metal.
2. Heated with strong sulphuric acid, they slowly blacken, and evolve a slight odor of burnt sugar.
3. Mixed in the cold with calcium chloride, in the presence of a slight excess of ammonium hydrate they give no precipitate; but on boiling, calcium citrate—$Ca_3(C_6H_5O_7)_2$—separates as a white precipitate. If this precipitate be filtered hot, and washed with a little boiling water, it is found to be quite insoluble in cold solution of potassium hydrate, but readily soluble in neutral solution of cupric chloride.

4. Mixed with argentic nitrate and boiled, no mirror of metallic silver is produced.

72. SEPARATION OF OXALATES, TARTRATES, CITRATES, and MALATES.

(Practise upon mixed solutions oxalic, tartaric, and citric acids.)

If the solution be acid, neutralise it with sodium hydrate; but if neutral or alkaline it is ready for use, and is treated as follows:—

A. Acidulate slightly but distinctly with acetic acid, bring the whole to the boil, and add a drop or two of calcium chloride; and if it produce a precipitate, add it till precipitation ceases. Keep the whole nearly boiling for a time, till the precipitate aggregates, and filter. The precipitate, after washing, should be quite insoluble in cold solution of potassium hydrate.

B. To the filtrate from (*A*) mixed with some more calcium chloride, ammonium hydrate is added in slight but distinct excess, and the whole thoroughly cooled. Calcium tartrate precipitates, and when it has settled clear, the liquid is poured off and preserved for (*C*). This precipitate, after washing, should be entirely soluble in cold solution of potassium hydrate, and reprecipitable by boiling.

C. The liquid is slowly boiled for some time; and if a precipitate does not form readily, a little more $CaCl_2$ and NH_4HO added, and the boiling resumed. The precipitate, when it begins to subside well, is filtered out whilst still hot. It should be (after washing) quite insoluble in cold solution of potassium hydrate; but soluble in perfectly neutral solution of cupric chloride.

D. To the filtrate add alcohol, when calcium malate will separate; but this portion of the separation is not infallible, and the precipitate must be carefully examined to see that it really is malate.

73, MECONIC ACID and MECONATES.

Meconic Acid is a white powder, with a strongly acid reaction, soluble in water, alcohol, and ether, and giving the reaction for meconates.

Meconates communicate a red color to ferric chloride solution. This color is *not discharged by $HgCl_2$ nor by dilute HCl.* By this means it is distinguished from a *sulphocyanate* and an *acetate.*

74. CARBOLIC ACID (or Phenol) and CARBOLATES (Phenates).

The qualities of this body are very distinctive.

1. It is a colorless, crystalline solid, melting at 94° F., and not volatile at 212°, having the odor and taste of creasote, being very poisonous, and not reddening test-paper.
2. The crystals deliquesce readily, forming a liquid which does not mix freely with water, but incorporates readily with alcohol, ether, and glycerine.
3. Mixed with HCl and exposed to the air on a strip of deal, it becomes greenish blue.
4. It coagulates *albumen.* It does not rotate polarised light.

5. Saturated with ammonia gas—NH_3—and heated in a closed tube, *aniline* is formed :—

$$C_6H_5HO + NH_3 = C_6H_5H_2N + H_2O.$$

6. It does not decompose carbonates.
7. NH_4HO and $CaOCl_2$ produce a blue liquid.
8. It unites directly with strong H_2SO_4 to form sulpho-phenic (or sulpho-carbolic) acid—$C_6H_5HSO_4$.
9. With bromine water it gives a white precipitate of tribromophenol—$C_6H_3Br_3O$.

Carbolates give the following reactions :—

1. When heated alone, they evolve the odor of carbolic acid and decompose.
2. Heated with strong sulphuric acid, they also smell of carbolic acid.
3. **Ferric chloride** causes a reddish-violet color.

Sulpho-Carbolates behave similarly, but also give the reactions of a sulphate with barium chloride.

75. BENZOIC ACID and BENZOATES.

Benzoic Acid is of characteristic appearance, being usually seen in light, feathery, flexible, nearly colorless crystalline plates or needles, and containing a trace of an agreeable volatile oil, unless it is the artificial acid prepared from naphthalene, which is odorless.

1. It is only slightly soluble in water, but dissolves freely in spirit, and in solutions of soluble hydrates.
2. Heated in the air, it burns with a luminous smoky flame ; and when made hot in a tube open at both ends, sublimes in needles, giving off an *irritating vapor*.

Benzoates possess the following general qualities :—

1. Heated with sulphuric acid they evolve the odor of benzoic acid, and darken.
2. Ferric chloride, in a solution made *slightly* alkaline by ammonium hydrate, gives a reddish-white precipitate—ferric benzoate—$Fe_2(C_7H_5O_2)_6$—soluble in acids (*benzoic* included). If this precipitate be now filtered out and digested in ammonium hydrate, it is decomposed into a precipitate of ferric hydrate, and a solution of ammonium benzoate, which is separated by filtration and treated as in 3.
3. Strong hot solutions of benzoates yield crystals of *benzoic acid* when hydrochloric acid is added and the solution allowed to cool.

76. SALICYLIC ACID ($HC_7H_5O_3$).

This acid occurs in prisms, when crystallised from a solution in alcohol in which it is readily soluble. It is freely dissolved by hot water, but not readily by cold, requiring 1,800 parts of the latter to completely dissolve it.

1. Its aqueous solution gives with Fe_2Cl_6 a *deep violet coloration*. The compounds with methyl, ethyl, etc., give this reaction, as well as the ordinary salts.
2. Its methyl salt, formed by warming a salicylate with sulphuric acid and wood spirit, has a very fragrant odor.

From most other solid bodies it may be separated by taking advantage of its exceptionally great solubility in ether. In the event of its presence in an organic liquid (such as milk) it or its salts may be procured in a pure condition by dialysis. The liquid is placed in a dialyser, and this floated on pure water. The salicylic acid or salicylate is found after some time to have entirely passed through the septum, and may be obtained by evaporating the aqueous solution and, when reduced in bulk, shaking with ether.

77. DETECTION OF CARBOLIC ACID IN THE PRESENCE OF SALICYLIC ACID.

Boil 10 grains in half an ounce of water, cool, decant the solution and add to it 1 minim of a saturated solution of $KHCO_3$, 1 minim of aniline, and 5 drops of solution of *chlorinated lime*, when, if carbolic acid be present, a deep blue is produced.

78. TANNIC, GALLIC, and PYROGALLIC ACIDS.

Tannic Acid is soluble in water and alcohol, and very soluble in glycerine. It is insoluble in pure dry ether, but dissolves readily in ether containing a little water.

Gallic Acid is slightly soluble in cold water, but readily in boiling; it is also freely soluble in glycerine, and slightly in alcohol and ether.

Pyrogallic Acid is very soluble in water, the solution rapidly absorbing oxygen from the air and becoming brown. It also dissolves in alcohol and ether.

DISTINCTION BETWEEN GALLIC, TANNIC, AND PYROGALLIC ACIDS.

Behaviour of the Acid with	Gallic.	Tannic.	Pyrogallic.
Ferrous salts . .	A dark solution is formed, gradually depositing a precipitate.	The same effect as Gallic.	A blue solution.
Ferric salts .	Purplish precipitate immediately formed.	Same as preceding.	A red solution.
Calcium hydrate in the form of *Milk of Lime.*	A brownish precipitate, becoming deep brown in a few seconds.	A white precipitate slowly changing.	Instantaneous production of a purple solution becoming brown by oxidation.
Gelatine . . .	No precipitate (except in the presence of gum).	Immediate brownish precipitate.	No precipitate.

CHAPTER IV.

QUALITATIVE ANALYSIS, AS APPLIED TO THE DETECTION OF UNKNOWN SALTS.

---

§ I. GENERAL PRELIMINARY EXAMINATION.

Under this head are included,—

1. The observation of the physical properties of the substance submitted for analysis.
2. Its behaviour when heated, either alone or in the presence of reducing agents or fluxes.
3. Its reaction with test-papers; the color it communicates to flame, etc.

So particular and minute may this examination be, that in the larger works on chemical analysis many pages will be found devoted to it; but for the purposes of the analysis likely to come before the ordinary chemical student, it is sufficient only to carry it the length of a few readily obtainable and unmistakable inferences.

Step 1. If the substance given for analysis be a liquid, carefully mark its reaction with blue and red litmus paper, evaporate a little to dryness at a gentle heat on a clean porcelain crucible lid, observing the nature of the residue left, if any; and finally raise this residue to a red heat, carefully noting whether it is volatilised, blackened, or altered in color any way. If a solid, heat it directly to redness on a crucible lid and note effect; then shake a little up with distilled water and note its reaction with blue and red litmus paper.

From a careful study of these points, the following simple inferences may safely be drawn; *any appearance not herein referred to being neglected as not affording a really distinctive indication*:—

A. **Neutral**, no odor, and leaving **no residue** whatever. Probably water.

B. **Strongly acid**, leaving **no residue.** Probably an ordinary volatile acid, such as HCl, HNO_3, $HC_2H_3O_2$, etc.

C. **Strongly acid**, leaving **a residue,** fusible by heat and also strongly acid. Probably a non-volatile mineral acid, such as H_3PO_4.

D. **Strongly acid**, leaving **a residue,** which on heating **chars,** and entirely burns away. Probably free organic acid, such as $H_2C_4H\ O_6$, $H_3C_6H_5O_7$, $HC_7H_5O_2$, etc.

Note.—Oxalic and formic acids do not char.

E. **Neutral or slightly acid**, leaving **a residue,** which volatilises in fumes, but without blackening. Probably an ordinary salt of a volatile metal, such as NH_4, Hg, As, Sb, etc.

F. **Neutral or slightly acid,** leaving **a residue** which on heating blackens and volatilises in fumes. Probably an organic salt of NH_4, Hg, or other volatile metal.

Note.—In this case it is best at once to test the original for NH_4 or Hg by boiling with KHO and $SnCl_2$ respectively.

G. **Neutral or slightly acid,** leaving **a residue,** which on heating changes color as follows:—

Yellow while hot, white on cooling. Probably salt of Zn.

Deep yellow while hot, yellow on cooling. Probably salt of Pb.

Yellowish-brown while hot, dirty light-yellow on cooling. Probably salt of Sn^{iv}.

Orange or red while hot, dull-yellow on cooling. Probably salt of Bi.

Red while hot, reddish-brown on cooling. Probably salt of Fe or Ce.

Permanent brownish-black. Certain salts of Mn.

H. **Neutral or slightly acid,** leaving **a white residue,** which blackens on heating, burns, and leaves a black or greyish mass. Probably an organic salt of a fixed metal. *In this special case, proceed as follows:—*

Moisten the residue with a little water, and touch it with reddened litmus paper. If alkaline, the original substance was an organic salt of K, Na, or Li, in which case proceed by (*a*). If not alkaline, proceed by (*b*).

(*a*) Boil the ash with the smallest possible quantity of water, filter, acidulate with HCl till effervescence ceases; dip a perfectly clean platinum wire in the solution, and try the flame test. If crimson, Li. Bright yellow, Na. Violet, K.

Note.—The latter flame not being very easily seen in daylight, it is advisable to add to the solution $PtCl_4$ and C_2H_6O. Shake well and cool. Yellow crystalline precipitate of potassium platinochloride ($PtCl_4 2KCl$).

(*b*) The ash is covered with water and treated with $HC_2H_3O_2$. If effervescence takes place, the original substance was probably an organic salt of Ba, Sr, or Ca; and these metals may at once be tested for in the acetic acid solution.

Note.—Oxalates, although organic, do not blacken to any extent. If carefully observed, however, a slight greyish tint, followed by a distinct glow running through the mass, will be noticed at the moment of decomposition. To make certain, it is well to place a little of the original powder in one tube, and the residue, after ignition, in another; cover them both with water, and add a drop of acetic acid to each. If the residue effervesce, and the original powder does not, strong presumptive evidence is obtained of the presence of an oxalate of the alkaline or earthy metals.

I. **Neutral or slightly acid,** leaving **a residue,** which takes fire and continues to burn even after removal from the flame, giving off clouds of white fumes and leaving a fixed white or pinkish residue. Probably a hypophosphite; which fact should be noted as an aid to future information as to acidulous radicals.

K. **Strongly alkaline,** leaving a fixed white residue, **also alkaline.** A hydrate, carbonate, bicarbonate, or sulphide of a fixed alkaline metal, or a hydrate or oxide of Ba, Sr, or Ca. In this case proceed as follows:—

Acidulate a portion of the original solution with HCl.

(*a*) If it effervesces without smell, and is therefore a carbonate or bicarbonate, test at once by the flame for K, Li, Na, and also another portion of the original solution with $HgCl_2$. If red, a carbonate; if not, a bicarbonate.

(*b*) Effervesces with smell of H_2S. In this case it is a sulphide; and if a deposit of S also takes place, a polysulphide. Add to a fresh portion of the original solution excess of HCl, boil till H_2S is expelled, filter, if necessary, and test the solution for all metals of fourth and fifth groups.

(*c*) Effervesces with smell of HCN. Probably an alkaline cyanide such as KCN.

(*d*) It does not effervesce. In this case add to a fresh portion of the original solution, $AgNO_3$. If a brownish-black precipitate be formed, it is a soluble hydrate. A portion of the original solution should be neutralised with HCl, and then examined for all metals of fourth and fifth groups.

Note.—If $AgNO_3$ with original solution gives a yellow, a white, or a brick-red precipitate, the presence of a phosphate, borate, or arseniate of K or Na may be suspected. In the case of a complex solution in which a salt of some other metal is given dissolved in excess of an alkali, an intimation of the fact will be obtained on cautiously adding the HCl, as, at the moment of neutralisation, the dissolved substance appears as a precipitate before again dissolving in the excess of HCl. Basic plumbic acetate has an alkaline reaction.

Step 2. Dip a clean platinum wire in the solution, or if a solid moisten the wire with HCl, dip it in the powdered substance, and heat in the inner Bunsen or blowpipe flame. The outer flame is colored as under :—

Violet . . .	Potassium.
Golden-yellow .	Sodium.
Yellowish-green .	Barium.
Crimson . .	Strontium or Lithium.
Orange-red . .	Calcium.
Green . . .	Copper or Boracic acid.
Blue . . .	Lead, Arsenic, Bismuth ; also Copper as chloride.

Step 3. Heat a little of the solid substance (or the residue left on evaporation if in solution) on charcoal before the blowpipe.

Ordinary alkaline salts fuse and sink into the charcoal ; some decrepitating (example NaCl, etc.), others deflagrating (as KNO_3, $KClO_3$, etc.), but no sufficiently characteristic indications are usually obtained, except in one of the following cases :—

A. A white luminous residue is left. Moisten it when cold with a drop or two of cobaltous nitrate, and again apply the blowpipe, observing any change of color as follows :—

The residue becomes blue, indicating Al, Silicates, or Phosphates.
" " " green, " Zn.
" " " pink or flesh-colored, indicating Mg.

B. A colored residue is left. Prepare a borax bead, and heat a little of the substance in it, both in the reducing and oxidising flame, and note any colors corresponding with the following list.

Metal.	In Oxidising Flame.	In Reducing Flame.
Cu	Green (hot). Blue (cold).	Red (cold).
Co	Blue.	Blue.
Cr	Green.	Green.
Fe	Red (hot). Yellowish (cold).	Bottle-green.
Mn	Amethyst.	Colorless.
Ni	Reddish-brown (hot). Yellow (cold).	Same as oxidising flame.

C. A metallic residue is left, with or without incrustation surrounding it. Mix a little of the substance with KCy and Na_2CO_3, and expose on charcoal to the reducing charcoal flame.

(*a*) Metallic globules are produced without any surrounding incrustation of oxide. This occurs with Ag, Au, Cu, Fe, Co, and Ni, all easily recognisable.

(*b*) Metallic globules are produced with a surrounding incrustation of oxide. This occurs with Sn, Bi, Pb, and Sb; the incrustation having the characteristic colors already described in Case I., Step 1, *G*.

Note.—Sb often forms a white and distinctly crystalline crust.

(*c*) The metal volatilises, and only leaves an incrustation of oxide. This occurs with As (garlic smell and white incrustation), Zn (yellow [hot], white [cold]), and Cd (reddish-brown).

§ II. DETECTION OF THE METAL PRESENT IN ANY SIMPLE SALT.

Step 1. Preparation of the solution for analysis for the metal, if the substance be not already dissolved.

1. Take a minute portion of the substance and boil it with water in a test-tube; should it dissolve, then take a large portion and dissolve for testing.
2. Should the salt prove insoluble, take another small portion and heat with HCl, and add a little water and again heat. If it now dissolves, prepare a larger quantity of the solution for use in the same manner.
3. Should it resist HCl, try another small portion with HNO_3 by heating, and then adding water. If this dissolves it, make up a larger quantity of a similar solution for testing.
4. Should HNO_3 also fail, try another small portion with two parts HCl and one part HNO_3, warming and diluting as before; and if it succeeds, make up a larger amount of solution in the same manner.
5. If all acids fail, then take another portion of the substance, mix it with several times its bulk of a perfectly dry mixture of sodium and potassium carbonates (prepared by heating Rochelle salt in an open crucible until the residue thoroughly ceases to evolve any gases, then extracting with distilled water, filtering, evaporating to dryness, heating the residue to redness, and preserving for use in a stoppered bottle. This reagent will hereafter be shortly described as **fusion mixture**). Place the whole in a platinum crucible, and fuse at a bright-red heat; when cold, boil with water and save the solution thus obtained for acidulous radicals. The insoluble matter is then to be drenched with strong HCl, slightly diluted and boiled, and the solution used for testing for the metal. Any insoluble white gritty matter still remaining is put down as silica.

Step 2. Detection of the metal.

The processes to be applied vary according to the limitation of the possible substances under examination, and the following tables are to be used accordingly, using the solution obtained in Step 1. *Remember that even when we have apparently found out the metal by the table we should always proceed to perfect confirmation by applying (to fresh portions of the solution each time) all the tests for the metal given in Chapter II.* Unless otherwise directed, all confirmations referred to in the tables are intended to be tried upon fresh portions of the original solution. For brevity the said solution is in the tables indicated by a capital **O** in thick type, and the word precipitate is contracted to ppt. In simple salts we go through the groups until we get a result, and as soon as we do so we stop and go no farther with the groups, but simply confirm the result obtained by special tests.

FULL TABLE FOR THE DETECTION OF THE METAL IN A SOLUTION CONTAINING ONE BASE ONLY.

N.B.—In using this Table, as soon as the Metal is discovered we go no further, and all confirmations, except where specially stated, are to be applied to fresh portions of the original Solution, hereafter represented by O.

1st Group.—Add a drop of HCl, and if it produce a precipitate, add excess.

Precipitate

May be either
$PbCl_2$, Hg_2Cl_2, $AgCl$ } White.

Confirmations.

Let ppt. settle, and pour off supernatant liquid; wash once by decantation with *cold* water, and then proceed as follows:—

(1.) Add some warm H_2O to ppt. and boil: if it dissolves and the solution gives a yellow with K_2CrO_4 = **Pb.**

(2.) If ppt. insoluble in water, add to it excess of NH_4HO.

(*a*) The precipitate turns black = **Hg₂** (ous).

(*b*) The precipitate dissolves and is re-precipitated by HNO_3 = **Ag.**

2nd Group.—To solution in which a drop of HCl has failed to produce a ppt. add H_2S, and if a discoloration appear, add excess and warm.

Precipitate

May be either
HgS, PbS, Bi_2S_3, CuS, CdS, As_2S_3, Sb_2S_3, SnS, SnS_2, PtS_2 or Au_2S_3

Let the precipitate settle, pour off as much of the supernatant liquid as possible, make the remainder alkaline by NH_4HO and add a few drops of NH_4HS and warm.

Insoluble.	Soluble.
CdS—yellow	As_2S_3, SnS_2 } Yellow.
HgS, PbS, Bi_2S_3, CuS } Blackish	Sb_2S_3—Orange
	SnS, PtS_2, Au_2S_3 } Blksh.

Confirmations.

(1.) *If ppt. yellow and insoluble in NH_4HS.* = Cd.

(2.) *If ppt. black and insoluble in NH_4HS.*

(*a*) Add to O, H_2SO_4: white = **Pb.**

(*b*) Add to O, KHO: yellow = **Hg**; white = **Bi**; and blue = Cu.

(3.) *If ppt. yellow and soluble in NH_4HS.*

(*a*) Add to O, KHO: white (sol. in excess and not re-precipitated on boiling) = **Sn** (ic).

(*b*) Add to O, KHO and Zn, boil and hold paper moistened with $AgNO_3$ over the tube: black stain = **As.**

(4.) *If ppt. orange, soluble in NH_4HS,* and re-precipitated orange by HCl = **Sb.**

(5.) *If ppt. brown or black, and soluble in NH_4HS.*

(*a*) Add to O, $HgCl_2$ and boil: grey = **Sn** (ous.)

(*b*) Add to O, $SnCl_2$: purple = **Au.**

(*c*) Add to O, KCl and rectified spirit: yellow = **Pt.**

3rd Group (Division A).—Evaporate a portion of original solution to dryness, and heat residue till any organic matter is destroyed; then dissolve in a few drops of HCl with the addition of a drop or two of HNO_3, dilute, and add NH_4Cl, and then NH_4HO in slight but distinct excess, and heat to boiling.

***Precipitate**

May be either
$Fe_2 6HO$ = red-brown.
$Cr_2 6HO$ = green.
Ce2HO, $Al_2 6HO$, Insol. phosph. of Fe, Cr, Al, Mn, Ba, Sr, Ca, and Mg. } White.

Confirmations.

(I.) *If ppt. red.*

(*a*) Test O with $K_6Fe_2Cy_{12}$: dark blue = **Fe.**

(*b*) Test O with K_4FeCy_6: dark blue = **Fe₂.**

(II.) *If ppt. green.*
Evap. portion of O on a porcelain crucible lid, add some KHO and a crystal of KNO_3, again evap. to dryness and fuse: yellow residue, sol. in H_2O and precipitated yellow by Pb2Ac after acidulation by HAc = **Cr.**

(III.) *If ppt. white.*

(1.) Add to O, KHO; white sol. in excess and re-precipitated by boiling with NH_4Cl = **Al.**

(2.) *If ppt. with KHO be insoluble in excess,* test O with $(NH_4)_2MoO_4$ dissolved in HNO_3 and boil.

(*a*) *If no precipitate forms,* test O for **Ce** by evaporating and igniting and getting a red residue, which dissolves with difficulty in strong HCl, and the solution diluted and nearly neutralized gives a white with $(NH_4)_2C_2O_4$.

(*b*) *If a yellow precipitate forms,* phosphoric acid is present, in which case proceed as follows:—Add to O, NH_4HO in excess, and then HAc until the solution is acid and boil.

(A) *A ppt. forms,* it may be $FePO_4$, $CrPO_4$, or $AlPO_4$, filter, wash, and treat on filter with boiling dilute KHO.

* *A reddish residue is left on filter* = **Fe** (test O for Fe and Fe₂.

† *A greenish residue is left on filter* = **Cr** (fuse as above.)

‡ *The ppt. dissolves* = **Al** (confirm by boiling filtrate with NH_4Cl.)

(B) *No precipitate forms.* We may have in solution the phosphates of Ba, Sr, Ca, Mn, or Mg. Test the same solution with K_2CrO_4: yellow = **Ba.** If no effect add a drop or two of dilute H_2SO_4, and stand for a short time: white = **Sr.** If not Sr add $(NH_4)_2C_2O_4$: white = **Ca.** If still not Ca cool and add excess of NH_4HO: white = **Mg** or **Mn.** Fuse ppt. on porcelain with KHO and KNO_3: green residue = **Mn.** If no green then the ppt. was **Mg.**

3rd Group (Division B).—To the solution in which NH_4HO has failed to produce a precipitate, add NH_4HS, and warm for some time.

Precipitate

May be either
MnS—flesh colrd.
ZnS—white.
CoS, NiS } black.

Confirmations.

I. *If ppt. white or flesh-coloured.*

(1.) Test O by evaporating & fusing with KHO and KNO_3: green residue = **Mn.**

(2.) Test O with K_4FeCy_6: white = **Zn.**

II. *If ppt. be black,* boil O with KHO.

(*a*) Blueish ppt. turning pinkish on boiling—**Co.**

(*b*) Greenish ppt. unaltered by boiling = **Ni.**

(Both should be confirmed by blowpipe bead.)

4th Group.—To same solution, add $(NH_4)_2CO_3$.

Precipitate

May be either
$BaCO_3$, $SrCO_3$, $CaCO_3$ } White

Confirmations.

Let ppt. settle, pour off as much supernatant liquor as possible, and dissolve in HAc, then test successively in the same liquid.

(1.) With K_2CrO_4: yellow = **Ba.**

(2.) With a drop of dilute H_2SO_4 and let it stand: white = **Sr.**

(3.) With $(NH_4)_2C_2O_4$: white = **Ca.**

5th Group.

(1). Divide into two portions; test one with Na_2HPO_4: white = **Mg.**

(2.) Evaporate the remainder to dryness, heat till fumes cease, dissolve residue in the smallest possible quantity of water, and try the flame test.

(*a*) Crimson flame = **Li.**

(*b*) Violet flame = **K.**

(*c*) Yellow flame = **Na.**

Confirm K by $PtCl_4$ and rectified spirit.

Test for NH_4 by boiling O with KHO and smelling.

* As Mn may sometimes precipitate through the solution absorbing oxygen, it is always advisable to filter out a little of this precipitate and fuse it with KHO and KNO_3:—green residue = **Mn.**

Table for the Detection of the Metal in a Simple Salt, the Metal present being limited to the Salts commonly used and included in the British Pharmacopœia.

(N.B.—*Ppt. means precipitate;* **O** *means the original solution.*)

1ST GROUP.—Add a drop of HCl, and if it produce a precipitate, add excess.

PRECIPITATE.

May contain—
Pb }
Hg (ous) } White.
Ag }

CONFIRMATIONS.

(1) Let ppt. settle; pour off liquid and add H_2O and boil. If it dissolves and gives a yellow ppt. with K_2CrO_4 = **Pb**.

(2) If not soluble in boiling H_2O, add NH_4HO. If now soluble = **Ag**. If turned black = **Hg** (ous).

2nd GROUP.—To the solution in which HCl has failed to give a ppt. add *excess* of H_2S water and warm gently.

PRECIPITATE.

(1) Notice its color and add to it a drop of NH_4HO to neutralise, and then add **10** drops of NH_4HS and warm. Notice whether it dissolves or remains insoluble. Thus we divide the group into two divisions, as follows:—

Division I.

(Insoluble in NH_4HS.)

Hg }
B }
Cu } = Black.
Pb }
Cd = Yellow.

If the ppt. was black, then to **O** add KHO.
Yellow = **Hg**.
Blue = **Cu**.
White (insoluble) = **Bi**.
White (soluble) = **Pb**.

Division II.

(Soluble in NH_4HS.)

Sb = Orange.
Sn(ous) = Brown.
As }
Sn(ic) } = Yellow.

Test **O**. (1) If ppt. brown:— With $HgCl_2$ for **Sn**(ous).

(2) If ppt. yellow; for **As** by Fleitman's test, and if not found then test specially for Sn(ic).

(3) If ppt. orange = **Sb**.

3rd GROUP.—To a *fresh* portion of **O** add some NH_4Cl, then NH_4HO in slight, but distinct excess, and observe effect; lastly, add a drop or two NH_4HS and again observe.

PRECIPITATE.

Division I.

A ppt. was caused by NH_4HO, and was afterwards turned black by NH_4HS. May be **Fe**, provided it it is readily bleached by HCl (*Ni and Co not so bleached*).
Test **O** for Fe(ous) with $K_6Fe_2Cy_{12}$.
" " Fe(ic) " $K_4Fe_2Cy_6$.

Division II.

A ppt. is caused by NH_4HO not altered in color by NH_4HS.
Green = **Cr** (*confirm by KHO*).
White = { Al. / Ce. / $Ca_3(PO_4)_2$.
To **O** add KHO.
White (soluble in excess) = **Al.**
White (insoluble in excess) = **Ce** or $Ca_3(PO_4)_2$.
Test specially for $Ca_3(PO_4)_2$ (see § V.)
(Ce salts always leave a red residue on heating).

Division III.

Ppt. with NH_4HO instantly soluble, but NH_4HS produced a light-colored ppt.
Flesh-colored = Mn.
White = Zn.
Add to **O** chlorine water and KHO.
Brown = **Mn.**
White (soluble) = **Zn.**

4th GROUP.—To the solution which has failed to give a ppt. in the 3rd group add $(NH_4)_2CO_3$.

PRECIPITATE.

May contain
Ba.
Ca.
(Sr).
To **O** (first neutralised *if acid* by NH_4HO, and then acidulated with acetic acid) add,
(1) K_2CrO_4 yellow ppt. = **Ba.**
(2) $(NH_4)_2C_2O_4$ white ppt. = **Ca.**

NOTE.—If the **O** gives a persistent red flame, **Sr** is to be suspected, and its absence must then be proved before testing for Ca (see p. 25).

5th GROUP.—To the solution in which $(NH_4)_2CO_3$ has failed to give a ppt. add Na_2HPO_4, cool and shake.

PRECIPITATE.

White = **Mg.**

6th GROUP.

—

(1) Test **O** for (NH_4) by boiling, with KHO and smelling for NH_3.

(2) Test **O** for K with $PtCl_4$, first concentrating by evaporation and cooling: yellow = **K.**

(3) Apply flame test.
Yellow = **Na.**
Crimson = **Li.**

§ III. DETECTION OF THE METALS IN COMPLEX MIXTURES OF TWO OR MORE SALTS.

Step I. Preparation of the solution for analysis in cases where the substance for analysis is not given in solution.

Note.—By carefully applying this step and intelligently judging the results, we can often reduce a separation of two salts to the performance of two separate simple analyses, and so save much time and trouble.

1. Boil some of the powdered substance with distilled water, and filter off from any insoluble matter.

   Evaporate a drop or two of the filtrate to dryness, at a gentle heat, on a slip of clean platinum foil, and if any residue be left, then save the balance of the filtrate for analysis as representing the portion of the original (if any) that is soluble in water.

2. If anything remains insoluble from water, then wash it on the filter with boiling water until a drop of the washings leaves no marked residue on evaporation. Rinse the insoluble off the paper into a tube, and add hydrochloric acid drop by drop (noting carefully any effervescence or odor as indicating the presence of certain acidulous radicals, such as carbonates, sulphides, sulphites, cyanides, etc.), and warm.

   If it now all dissolves, save the fluid for analysis. If not, then separate the insoluble, test the filtrate by evaporation of a drop or two, to see whether anything has dissolved, and if so, save the fluid for analysis as representing the metals present in the form of salts insoluble in water, but soluble in HCl.

   Note.—This division of any mixture into salts soluble and insoluble in water gives the greatest assistance in the subsequent testing for the acidulous radicals. For example, if a metal of the 5th group be found in the portion soluble in water, then any acidulous radical almost may be present; while if a metal of one of the other groups be found, then generally speaking only a nitrate, sulphate, chloride or acetate need be first searched for. If, on the other hand, the substance resists water and only goes into solution with HCl, then as a rule *no metal of the 5th group is present*, and we might consider that we were probably dealing with a carbonate, oxide, phosphate, arseniate, oxalate, sulphide, sulphite, cyanide, ferro- or ferri-cyanide, or borate of a metal, not in the 5th group. Certain tartrates and citrates, chiefly of the 4th group, would also come in this category.

3. If the substance resists both water and HCl, try nitric acid, first alone, and then with the addition of hydrochloric acid.

   This treatment dissolves certain metals in the free state, such as Ag, Pb, Bi, Hg, and Cu, and also acts upon HgCl, HgS, and other insoluble sulphides, and on Fe_2O_3 and some refractory oxides. Gold and platinum only dissolve in nitro-hydrochloric acid.

   Note.—When HNO_3 has been used as a solvent, the liquid should always be evaporated with HCl till all the HNO_3 has been displaced, then allowed to get quite cold and any precipitate filtered out and treated as belonging to the 1st group, while the filtrate is directly treated with H_2S.

4. If anything still remains insoluble, it must be fused with fusion mixture ($KNaCO_3$) at a bright red heat till action ceases, and the residue so obtained boiled with water and filtered.

   The filtrate is used for the detection of acidulous radicals; while the insoluble matter is dissolved (after washing) in nitric acid and used for finding the metals. The usual run of articles requiring this treatment are—sand, clay and other silicates, sulphates of Ba, Sr, Ca (latter not always) and Pb, the haloid salts of silver, SnO_2 and Sb_2O_3.

Step II. Proceed to apply the following tables to the prepared solution from Step I.

Note.—The whole of the first table for "separation into groups" must be gone through, but if no effect be obtained in Groups I. or II., a fresh portion of the prepared solution should be taken for Group III., etc., so saving the time required for evaporating to a considerable extent.

5

TABLE FOR THE SEPARATION OF METALS INTO GROUPS.

Add a drop of HCl and if it produce a precipitate, add excess (*Note* 1).

which may contain
$PbCl_2$
Hg_2Cl_2
$AgCl$,
is collected on a filter and examined by Table A

GROUP I.

To the filtrate add H_2S water, and if it produce a discoloration, warm to blood-heat and pass H_2S in excess.

The precipitate is to be carefully washed with boiling H_2O till quite free from HCl (*Note* 2). It is then to be washed into a test-tube about half-full of H_2O, and from 10 to 20 drops of NH_4HS having been added, the whole is warmed for a time and filtered (*Note* 5).

(*a*) Insoluble portion, including possibly
HgS
PbS
Bi_2S_3
CuS
CdS,
is examined by Table B.

(*b*) Soluble portion, containing possibly
As_2S_3
Sb_2S_3
SnS
SnS_2
Au_2S_3
PtS_2,
is examined by Table C.

GROUP II.

Evaporate filtrate to dryness in a platinum capsule, and heat till any organic matter present is decomposed. Moisten the residue with strong HCl, and then add H_2O and boil (any insoluble (*Note* 3) if white SiO_2). Take a little of the solution, add HNO_3 and $(NH_4)_2MoO_4$ and boil. If Phosphates be present, a yellow precipitate will be produced (*Note* 4). Take now the rest of the solution, warm it with a drop or two of HNO_3 and then add NH_4Cl and NH_4HO, the latter in slight but distinct excess; boil and filter (*a*). Add to the filtrate NH_4HS in excess, boil and filter (*b*).

(*a*) NH_4HO
Precipitate may (in absence of PO_4) contain
Fe_26HO
Cr_26HO
Al_26HO
$Ce2HO$
Wash with boiling H_2O and examine by Table D.
(In presence of PO_4)
It may contain
$FePO_4$
$AlPO_4$
$CrPO_4$
Ce_32PO_4
Ba_32PO_4
Sr_32PO_4
Ca_32PO_4
Mn_32PO_4
Mg_32PO_4
Wash with boiling water and examine by Table E.

(*b*) NH_4HS
Precipitate may contain
MnS
ZnS
NiS
CoS
Wash with H_2O containing NH_4HS and examined by Table F.

Division (a). GROUP III. Division (b).

Add to filtrate $(NH_4)_2CO_3$ and boil.

Precipitate may contain
$BaCO_3$
$SrCO_3$
$CaCO_3$
examined by Table G.

GROUP IV.

Filtrate may contain
Mg
Li
K
Na
examined by Table H.

GROUP V.

NOTES.

(1) Too much excess HCl prevents the rapid precipitation of Cd by H_2S.

(2) Known by testing washings with $AgNO_3$ and continuing to wash and test till no precipitate is produced.

(3) Cerium leaves a very dark red residue only soluble in strong acid.

(4) A little patience here, as in dilute solutions the precipitate does not form instantly.

(5) When Cu is suspected from the Preliminary Examination, take care not to use yellow NH_4HS as it always dissolves some CuS.

TABLE A.

SEPARATION OF METALS OF GROUP I.

After washing the precipitate once with cold water; put the funnel over a fresh tube and pour on some boiling water.

<table>
<tr>
<td rowspan="2">The filtrate may contain
$PbCl_2$
Test while still hot with K_2CrO_4
Yellow Precipitate = Pb.</td>
<td colspan="2">Any precipitate remaining on the filter is washed with boiling water till all trace of Pb is removed, and then percolated with dilute NH_4HO (1 in 3).</td>
</tr>
<tr>
<td>Any black precipitate which remains on the filter is NH_2Hg_2Cl, and proves the presence of Hg_2.</td>
<td>The filtrate is diluted with an equal bulk of water, and then cautiously acidulated with HNO_3 (to re-precipitate AgCl previously dissolved by the NH_4HO).
White precipitate = Ag.</td>
</tr>
</table>

NOTE.—Instead of filtering, the whole of this analysis may be done by decantation, as the precipitates are heavy and settle rapidly.

TABLE B.

SEPARATION OF METALS OF GROUP II., DIVISION (a).

Wash precipitate with boiling water, and then transfer it to a small porcelain dish, add a few drops of HNO_3, and warm till red fumes cease, and repeat this heating with HNO_3 till an additional drop fails to cause any more red fumes. Now wash the contents of the basin into a tube with a little water (1), add H_2SO_4 till no more precipitate forms, then cool and add an equal bulk of spirit of wine (methylated).

<table>
<tr>
<td rowspan="2">Any Precipitate may contain HgS (2), and $PbSO_4$. Percolate with a strong solution of $NH_4C_2H_3O_2$, and test filtrate with K_2CrO_4. Yellow = Pb.
Any black residue now remaining in the filter will be HgS, and is to be confirmed by dissolving in HCl by the aid of a crystal of $KClO_3$, and (after boiling free from Cl) adding $SnCl_2$, and boiling. Grey precipitate = Hg.</td>
<td colspan="2">To the Filtrate, which may still contain Bi, Cu, and Cd, add NH_4HO in excess.</td>
</tr>
<tr>
<td>Any Precipitate will be Bi3HO, and is to be confirmed by dissolving in the least possible quantity of HCl by the aid of heat, and pouring into H_2O.
White = Bi.</td>
<td>The Filtrate, if blue in colour, contains Cu for certain, and possibly also Cd. If not blue, no Cu is present, and Cd is to be directly tested for by adding NH_4HS. Yellow = Cd. If Cu be present, then add KCy until the blue colour is discharged, and pass H_2S, when any Cd will be precipitated as yellow CdS.</td>
</tr>
</table>

NOTE 1.—Too much water added might precipitate any Bi, as Oxy-Nitrate.

2.—If any Cl have been left in the original group precipitate (through inefficient washing) the HgS will dissolve and be lost.

TABLE C.

SEPARATION OF METALS OF GROUP II., DIVISION (b).

First Method, in Absence of Gold and Platinum.

Acidulate with HCl to cause the reprecipitation of the sulphides. If a decided yellow, orange, or black precipitate separates, then proceed to filter out; but if only a cloud of precipitated sulphur forms (not separating readily, and nearly white in colour), no metals of this division are present. After washing the sulphides, suspend them in a cold solution of the B.P. ammonium carbonate, shake up for a few minutes, and filter.

The **filtrate** may contain $(NH_4)_3AsS_3$. Add HCl in excess, and if present As will be reprecipitated as yellow As_2S_3. Filter out and confirm by drying and heating with KCy and Na_2CO_3 in a small tube, thus getting a mirror of metallic arsenic, or, if preferred, by Fleitman's test.

Any **orange** or **yellow precipitate** still remaining may be sulphides of Sb or Sn^(ic)^. Dissolve in strong boiling HCl, dilute a little, and put into a platinum dish with a rod of zinc so held that it dips in the fluid and touches the platinum outside the liquid. Electrolysis will set in and cause the Sb to deposit as a black stain closely adhering to the platinum, while Sn will deposit as loose metallic granules. Boil with HCl (after washing) and the Sn will dissolve, forming $SnCl_2$ and giving a precipitate with $HgCl_2$; while the Sb will remain undissolved, and may be confirmed by oxidizing with a drop of HNO_3, then dissolving in solution of $H_2C_4H_4O_6$ and getting the characteristic precipitate with H_2S.

Second Method in Presence of Gold and Platinum.

When the precipitate from the NH_4HS by dilute HCl is dark in colour, either Sn^(ous)^ Au or Pt must be present. In this case the separation had better be conducted as follows. Boil the mixed sulphides at once with strong HCl as long as H_2S is given off; dilute slightly, and filter.

Filtrate may contain $SbCl_3$ and $SnCl_2$, which are separated by electrolytic process above described.

Precipitate may contain As_2S_3, Au_2S_3, and PtS_2. If it be yellow, simply confirm the As by fusion in a tube as above; but if it be blackish, then Au or Pt are certainly present. In this case digest the precipitate with solution of ammonium carbonate as above.

Filtrate may contain $(NH_4)_3AsS_3$. Reprecipitate with HCl, and prove by fusion as above.

Precipitate may be Au_2S_3 and PtS_2. Dissolve in aqua regia; dilute, and test a portion for Au with $SnCl_2$ (Purple = Au), and another for Pt by KCl and S.V.R. Yellow = Pt.

TABLE D.

SEPARATION OF METALS OF GROUP III., DIVISION (a).

(*Note* 1.) Wash and transfer to a platinum or silver dish, and fuse with $KNaCO_3$ and KNO_3. Boil the residue in water and filter.

Insoluble portion	Solution	
Insoluble portion may contain Fe and Ce. (*Note* 2.) Moisten with strong HCl, and then add H_2O and boil. Test a portion of the solution for Fe by K_4FeCy_6 Blue = Fe. To remainder add citric acid, and then excess of NH_4HO, and finally $(NH_4)_2C_2O_4$. White = Ce.	The solution (*which if yellow contains* Cr) is mixed with Na_2HPO_4 acidulated with $HC_2H_3O_2$ and boiled.	
	Precipitate is $AlPO_4$. Gelatinous white = Al.	**Filtrate** treated with $AgNO_3$ gives a red, or with $BaCl_2$ a yellow if Cr be present (*Note* 3).

NOTES.

1. If this group precipitate be deep reddish-brown, nothing is probably present but Fe. If it be at all greenish, Cr is present. If it be white or pale red, take a little on a watch-glass, and touch it with an excess of KHO and observe effect.

   Originally white and soluble, only Al.
   ,, ,, insoluble ,, Ce.
   ,, pale red, and becoming deep red, and partly dissolving, Fe and Al both present (probably).
   ,, ,, ,, not darkened by KHO, but darkened by NaClO, Fe and Ce both present.

2. If Ce be present, the red residue is with great difficulty soluble, even in strong HCl.
3. Sometimes the NH_4HO brings down Mn along with the Fe in this group; and both the residue and solution being then bright green, the yellow of the Cr is of course masked. In this case boil the filtrate with HCl and S.V.R., when the green of the manganate will be removed, while the chromate will be reduced to Cr_2Cl_6 and turned greenish. If therefore the originally green solution be entirely decoloured by HCl and S.V.R. nothing but Mn was present; but if a green tint remain, then Cr was also present. This is of rare occurrence, but must always be kept in view when the original fusion gives a green residue.

TABLE E.

SEPARATION OF METALS OF GROUP III. (DIVISION a) IN PRESENCE OF PHOSPHORIC ACID.

(*Note* 1.) Dissolve precipitate in dilute HCl, add excess of Na_2HPO_4 and then excess of $NH_4C_2H_3O_2$ and boil.

<table>
<tr>
<td colspan="2">Precipitate may be (Note 2) $FePO_4$, $AlPO_4$, Ce_32PO_4 wash with boiling H_2O and percolate with some boiling dilute KHO.</td>
<td colspan="4">Filtrate, if green in colour, contains Cr. (but if perfectly colourless, that metal may be noted as absent). Add K_2CrO_4 and warm.</td>
</tr>
<tr>
<td rowspan="3">Residue may contain Fe and Ce
Dissolve in HCl and test a portion for Fe by K_4FeCy_6.
Blue = Fe.
To the remainder add citric acid and then NH_4HO in excess. An immediate white or a white after adding $(NH_4)_2C_2O_4$ = Ce.</td>
<td rowspan="3">Filtrate may contain $AlPO_4$
supersaturate with $HC_2H_3O_2$.
White gelatinous = Al.</td>
<td rowspan="3">Precipitate
$BaCrO_4$.
Yellow = Ba.</td>
<td colspan="3">Filtrate must be tested for Sr by flame test (If present, Note 3), if absent, proceed as follows. Add $(NH_4)_2C_2O_4$.</td>
</tr>
<tr>
<td rowspan="2">Precipitate
is CaC_2O_4.
White = Ca.</td>
<td colspan="2">Filtrate is mixed with a little $H_3C_6H_5O_7$ cooled and excess of NH_4HO added (Note 4).</td>
</tr>
<tr>
<td>Precipitate may contain
Mn_32PO_4 or Mg_32PO_4
(See Note 5).
Separation, if necessary, by dissolving in dilute HCl, adding excess of $NH_4C_2H_3O_2$, boiling, and dropping in Fe_2Cl_6 till a reddish tint is produced. Filter out the $FePO_4$ thus produced, and to the filtrate add NH_4Cl, NH_4HO, and NH_4HS, to remove Mn and Fe. Again filter, concentrate to a small bulk, and test when quite cold for Mg by NH_4HO and Na_2HPO_4.</td>
<td>Filtrate
may contain Cr only, prove by fusion if necessary.</td>
</tr>
</table>

NOTES.

1. This precipitate, if free from Ce, Cr, and Mn, is much more simply worked. Therefore take a little before dissolving and put it on a porcelain crucible-lid with a few drops of strong KHO and a crystal of KNO_3 evaporate to dryness and fuse. If no bright yellow or green residue be formed Cr and Mn are respectively absent. Take another portion of the group precipitate on a white porcelain lid, and add some solution of chlorinated soda or lime; if it does not turn yellow, then Ce is absent (if Mn be present this last test will turn the precipitate brown) so that it is only when it does not change colour at all that the positive absence of Ce can be insured.
2. Here $CrPO_4$ should be found, as it is theoretically insoluble in acetic acid; but in practice it is, when freshly precipitated, nearly entirely soluble, and passes on as shown in the Table. If its presence is suspected, a little of the precipitate may be fused as above.
3. If Sr be present (known by its crimson flame, and distinguished from the mere orange-red of Ca) it must be removed by adding a little very dilute H_2SO_4, letting stand for twenty minutes in a warm place, and then filtering out the $SrSO_4$ formed. In this case excess of ammonium acetate must be added before proceeding to test for Ca.
4. If no Cr is present the citric acid need not be used, as it is only added to prevent the precipitation of $CrPO_4$ by the NH_4HO.
5. If no Mn was found by the preliminary fusion (*Note* 1) this precipitate is all Mg; but if Mn be present it must be first removed before Mg can be definitely proved.

TABLE F.

SEPARATION OF METALS OF GROUP III., DIVISION (b).

Case I.—If the group precipitate be white or flesh-coloured, nothing is present but MnS and ZnS; and it then only remains to dissolve in dilute HCl, add NaHO in excess, boil, and filter.

The **filtrate** may contain Zn dissolved in excess of NaHO; add excess of $HC_2H_3O_2$ and then K_4FeCy_6. White = Zn.	Any **precipitate** may be Mn2HO, which must be confirmed by fusion on platinum foil with $KNaCO_3$ and KNO_3. Green residue = Mn. (*Note.*) The supposed precipitate of Mn must be confirmed, because any trace of Fe escaping precipitation in its own group might appear here as Mn.

Case II.—If the group precipitate be dark in colour, then Co and Ni may also be present. In this case a black residue of NiS and CoS will remain insoluble in the diluted HCl. Any such residue must be filtered out and dissolved in strong boiling HCl, or Aqua Regia, if necessary; then add excess of KCy, and boil in the fume chamber, adding more HCl if necessary, until all further smell of HCy ceases to be developed. Then add KHO in excess.

Any **precipitate** contains Ni; confirm by drying and heating in borax bead.	The **filtrate** may contain Co; confirm by evaporating to dryness, and heating residue in borax bead.

TABLE G.

SEPARATION OF METALS OF GROUP IV.

Dissolve the precipitate in $HC_2H_3O_2$ and add K_2CrO_4 in excess.

<table>
<tr>
<td rowspan="2">Any precipitate is
$BaCrO_4$. Yellow = Ba.</td>
<td colspan="2">Take a small portion of the filtrate and add some saturated solution of calcium sulphate, warming gently. If no precipitate should form, Ca only is present, and may be confirmed at once by adding $(NH_4)_2C_2O_4$ to the remainder of the filtrate. But if a precipitate be produced Sr is present, and must be separated by adding a few drops of very dilute H_2SO to the said remainder, and letting the whole stand in a warm place for fifteen or twenty minutes.</td>
</tr>
<tr>
<td>Any precipitate is $SrSO_4$; confirmed by being perfectly insoluble after digestion for some time at a gentle heat with a strong solution of $(NH_4)_2SO_4$ and a little NH_4HO (to remove any $CaSO_4$ accidentally precipitated). White residue giving crimson flame = Sr.</td>
<td>The filtrate may contain Ca, confirmed by adding NH_4HO and $(NH_4)_2C_2O_4$. White = Ca.
(Note.) To ensure that Sr is fully separated, see that a portion of this filtrate gives no precipitate by warming with a saturated solution of Ca SO_4.</td>
</tr>
</table>

TABLE H.

SEPARATION OF METALS OF GROUP V.

Divide the solution into $\frac{1}{3}$ and $\frac{2}{3}$. Take the $\frac{1}{3}$ portion, and concentrate by evaporation if very dilute, cool perfectly, and add excess of NH_4HO and Na_2HPO_4. White = **Mg**. Evaporate the filtrate to dryness on the water bath, and take up with *boiling water*, when any white insoluble residue is **Li**, provided it gives a crimson flame on a platinum wire previously dipped in HCl.

Take the remaining $\frac{2}{3}$ and evaporate to dryness, and heat till all fumes of ammonium salts cease to be evolved. Take up the residue with the smallest quantity of boiling water (pouring off from and disregarding any insoluble matter). Then acidulate with HCl, dip in a wire and test flame, looking first at it with the naked eye, and then through a piece of blue glass (to cut off the sodium yellow).

Red	Li (disregard if already found in the $\frac{1}{3}$ portion).
Yellow	Na.
Violet	K.

Confirm K by adding $PtCl_4$ to the rest of the acidulated solution, and getting when cold a yellow precipitate of $PtCl_4 2KCl$.

Test for NH_4 by boiling some of the original solution with KHO, and getting off NH_3.

§ IV. DETECTION OF THE ACIDULOUS RADICALS.

Division A.—Preliminary Examination.

IMPORTANT NOTE.—We must always decide what metals or bases are present before we proceed to test for acidulous radicals. We must then note which bases are present as *soluble* and which as *insoluble* salts (in H_2O). Lastly, we must consider what acids might be present in each case, and only test for such possible acids, because *nothing leads to so many errors as testing for acids which could not possibly exist.* We must also carefully note the information received in the former preliminary examination, especially as regards the presence of organic matter, and remember that, if the original substance does not char on heating, *we must never enter into the testing for organic acids,* because none can possibly be present except oxalate, which is provided for in the inorganic portion of the course with this very object, together with a few others included for convenience. We must also remember to note what happens when we dissolve the original substance in HCl, provided such a step is necessary, and if any effervescence occurs we must be sure to smell the gas given off, because we may then at once detect the following :—

*Carbonate . . effervescence without odor, and the evolved gas poured into lime-water renders it milky.
*Sulphide . . odor of H_2S (with deposit of S polysulphide).
*Sulphite . . " SO_2 (" " hyposulphite).
*Cyanide . . " HCN.
Peroxide . . " chlorine.
Fe, Zn, or Sn (as metals)—no odor but hydrogen evolved.

We must also remember that organic bodies, such as alkaloids or sugar, other than organic salts, might be contained in a mixture which would cause charring on heating, and so lead us to test for what was not there. It will be useful at this point to see how we can guard against two of the more commonly occurring of such cases.

(1) *Sugar.* This will cause the soluble portion to be syrupy, and when warmed with *dilute* H_2SO_4 it will rapidly darken, whereas organic salts, as a rule, require fairly strong H_2SO_4 to char them. The solution will have a sweet taste, and after boiling with a drop or two of very dilute H_2SO_4 it will reduce Fehling's solution.

(2) *Alkaloids* (nitrogenous organic bases). These will cause an odor like burning hair on heating to redness. The soluble portion of the mixture carefully treated with very dilute NH_4HO will usually give a cloud (which may or may not dissolve in excess) and then the same liquid shaken up with chloroform, and the chloroform evaporated at a *gentle* heat, will leave the alkaloid as a residue. If no residue be thus obtained then no alkaloid can be present except morphia, and this latter would never be put in a mixture unless specially intended for toxicological investigation, because

* In soluble salts these effects will come on adding HCl in Group I.

its detection requires altogether special work, which will be afterwards detailed.

Having well considered all this, we now proceed to the actual work, carefully remembering that all the indications are merely preliminary, and that we are *not to take notice unless we really get a distinct result.* If we really do get one, then it may save us going so far through our actual acid course, but if we are not certain, then it is no use attempting to persuade ourselves and wasting time, but we should just note the probability and then at once pass on to confirm by the actual course hereafter detailed.

No attempt is made to describe odors, because the student should simply put himself through a course of training for this preliminary examination on known salts, and learn to recognise all the odors, etc. This is a most important study, and should be carefully stuck to, until the nose and eyes have been quite trained to recognise the individual effects to be expected from each acid.

Step I. Put a portion of the original solution in a tube, or if it be a solid cover it with some water, *just acidulate* with *dilute* H_2SO_4 and look for any effervescence or odor, then boil and smell. The following radicals may be thus recognised :—

Effervescence without odor . . .	Carbonate.
Effervescence with characteristic odors .	Sulphide. Sulphite. Cyanide. Hypochlorite.
Red fumes .	Nitrite.

Step II. Add another drop of H_2SO_4, and again warm.

Odor of vinegar	Acetate.
„ „ SO_2 with deposit of S .	Hyposulphite.
„ „ HCN „ „ „ .	Sulphocyanate.
„ „ HCN „ crystalline deposit, often bluish	*Ferro-* or *Ferri-*cyanide.
„ „ Valerian or sharp odor . .	Valerianate, Benzoate, Succinate.
„ „ Carbolic acid . . .	Carbolate.

Note.—The effects of Step II. will often come perfectly in Step I., and then Step II. may be considered as part of Step I.

Step III. Put a little of the original solid (or the residue left on evaporation if the original was a liquid) into a dry tube, cover it with strong H_2SO_4, and warm, but not sufficiently to cause the H_2SO_4 itself to fume. (*See note, important to prevent accidents.*)

Thus we get :—

White fumes (characteristic of)	Chloride, Nitrate, Fluoride. Benzoate. Succinate. Sulpho-carbolate.	Change of color and colored fumes (characteristic of)	Iodide. Iodate. Bromide. Bromate. Chlorate.

Observation	Indication
Effervescence on warming only, which persists after withdrawing from flame, but with *no* darkening in color and no odor.	*Formates*—give off CO only, and consequently the gas does not affect lime-water. *Oxalates*—give both CO and CO_2, and the gas therefore renders lime-water milky.
Effervescence on warming, but the liquid darkens in color to a greater or less extent.	*Tartrates*—rapid charring and smell of burnt sugar. *Lactates*—not so dark, and peculiar odor. *Citrates*—slow darkening and peculiar sharp odor. *Oleates*—char and give odor of acrolein.
Darkening in color without any very marked effervescence.	Meconate. Tannate. Gallate Pyrogallate. Salicylate (*very* slow darkening).

No fumes—gelatinous deposit (or flaky)—Silicate.

" "—scaly crystals with pearly lustre—Borate (best seen on cooling).

No change takes place at all with—Sulphate, phosphate, and arseniate.

Chromates turn orange and then green—Bichromates turn green straight off.

NOTE IMPORTANT.—**On adding strong sulphuric acid to any solid, one drop only should be first carefully applied, because chlorates, iodates, etc., are apt to explode on the first touch of the acid.**

If we get a decided indication of the presence of any acidulous radical as above, we may at once apply confirmatory tests for the radical found to our original substance, and so save going through the course, especially if the substance be soluble in water; but if insoluble a solution must always be specially prepared for acid testing.

Division B.—Preparation of a Solution for Testing for Acidulous Radicals.

The success of the course for the detection of acids depends in the highest degree upon the care with which the solution is first prepared. It may be taken as a general rule that no testing for acids is reliable unless they are present in the form of salts of alkaline metals. It is therefore necessary to transform our acids into such salts; and, to do this successfully, the following rules must be closely adhered to:—

I. If the original is soluble in water, and *absolutely neutral* to test-paper, you may venture as a rule to use it as it is, and this will also apply, if it be *alkaline*, to test-paper.

II. If the original be soluble in water, but in the least acid, we must drop in NaHO till it is rendered just *alkaline*, boil, and, if any precipitate should form, filter and use the filtrate for the acid course.

III. The portion insoluble in water (or the whole of the original if all insoluble) must be boiled with a little NaHO, then diluted, filtered, and the filtrate only used for the acid course.

Note.—If Al, Zn, Sb, Sn, Pb, or any metal whose hydrate is soluble in excess of NaHO has been found, then we must use a solution of Na_2CO_3 instead of NaHO in both Cases II. and III.

We must also take care to prepare plenty of our solution, because if the full acid course has to be gone through, we shall require possibly to employ eight to ten different portions before we have finished.

This course now about to be explained is so devised that by working upon the prepared solution, in the presence successively of HCl, HNO_3, $HC_2H_3O_2$, H_2SO_4 and absolute neutrality, we can insure the precipitation in each stage of certain given acidulous radicals only by reagents which, if used without such precautions, would precipitate many more than they do when so employed.

Division C.—Course for the Detection of Inorganic Acids together with a few Organics included for certain reasons.

Step I. Acidulate a portion of the prepared solution with HCl, and then to successive portions thereof apply the following tests :—

Reagent.	Effect.	Acid Present.
(*a*) $BaCl_2$	White ppt. insoluble in boiling HNO_3	Sulphate.
(*b*) Fe_2Cl_6 . .	Dark blue ppt. . . .	Ferrocyanide.
	Blood red color discharged by $HgCl_2$ . .	Sulphocyanate.
	Blood red color not discharged by $HgCl_2$ . .	Meconate.
(*c*) $FeSO_4$ . .	Dark blue	Ferricyanide.
(*d*) Turmeric paper .	Dip in and dry over the gas when the paper turns pink	Borate.

Step II. Acidulate a portion of the prepared solution with HNO_3, add *excess* of $AgNO_3$, warm and shake, *disregarding any precipitate that is not white or yellow and distinctly curdy.* Thus we get the following :—

(*a*) Cyanide—Curdy white; soluble in very dilute NH_4HO, and also in boiling HNO_3.

(*b*) Chloride—Curdy white; soluble in very dilute NH_4HO, but insoluble in boiling HNO_3.

(*c*) Bromide—Curdy dirty white; slowly soluble in fairly strong NH_4HO, but not in very dilute; insoluble in HNO_3.

(*d*) Iodide—curdy pale yellow; insoluble even in strong NH_4HO and also in HNO_3.

Note.—Many other acids, such as ferrocyanide, oxalate, chromate, etc., are apt to come down with $AgNO_3$ in presence even of HNO_3, but the precipitates are (if white) *not curdy*, or they are colored red and so will be disregarded; and we therefore deal only with the four acids mentioned giving curdy precipitates.

To distinguish between these four acids we—

(1) Filter out the precipitate with $AgNO_3$, wash it, and then percolate it several times with very dilute NH_4HO (1 in 20), when AgCl and AgCN will dissolve, and can be reprecipitated from the filtrate by HNO_3, while any AgBr or AgI will be left on the filter.

Note.—It is very important to have the dilute NH_4HO exactly the correct strength, because, if stronger, then AgBr will also dissolve, and in any case a *mere cloud on adding the HNO_3 is to be disregarded*, because if AgCl or AgCN be really present, they will reprecipitate in distinct curds, on adding HNO_3, warming and shaking.

(2) If by (1) evidence of the presence of Cl or CN be obtained, then test a portion of the original prepared solution for CN by Scheele's test, and if not present then the precipitate was all due to Cl. If CN be found, then another precipitate must be obtained by excess of $AgNO_3$, filtered, washed, drained, and transferred to a tube with strong HNO_3 and boiled, when any AgCl will remain insoluble and also prove its presence.

Note.—As HCN is so easily smelt in the preliminary examination we should always know before we begin the group whether it is there, and then if it be present the boiling with HNO_3 will be required, but if not, then we put it down at once as chloride if the NH_4HO dissolves anything.

(3) If, after treating with NH_4HO (1 in 20), any residue be left on the filter, leading to the idea that AgBr or AgI may be present, we proceed as follows: To a *small* portion of our prepared solution a drop of mucilage of starch is added, and then one or two drops of chlorine water. If **iodide** be present we shall get a blue. Now we go on adding fresh chlorine water till all the blue has been bleached, and if the whole is now perfectly white only iodide is present; but if it remain at all yellow, then we add some chloroform and shake up, when an orange color in the chloroform will indicate **bromide**. This depends on the fact that free iodine combines with chlorine more readily than with bromine.

Step III. Acidulate a portion of the prepared solution with acetic acid, bring it to the boil, and then test successive portions while boiling as follows:—

REAGENT.	EFFECT.	ACID PRESENT.
(*a*) $CaCl_2$ . . .	White ppt. soluble in HCl	Oxalate.
(*b*) Fe_2Cl_6 (*not* in excess) .	White ppt.	Phosphate or Arseniate.
(*c*) $Pb(C_2H_3O_2)_2$ . .	Yellow	Chromate.

To distinguish between phosphate and arseniate *exactly* neutralise a portion of the prepared solution with dilute HNO_3 and add $AgNO_3$.

Yellow soluble in NH_4HO = Phosphate
Red do. do. = Arseniate

Step IV. *Just acidulate* a portion of the prepared solution with dilute H_2SO_4, then add a strong and fresh solution of $FeSO_4$ and run some strong H_2SO down the side of the tube so that it collects at the bottom. A dark ring where the liquids meet proves **Nitrate.**

Note.—If *iodide* has been previously found, this test fails to be conclusive, and in such case we must take advantage of the power of nascent hydrogen to reduce nitrates to ammonia. If no salt of NH_4 has been found in metal testing, we add to some of the prepared solution a fragment of zinc and sufficient HCl to cause a brisk effervescence. After ten minutes we add excess of KHO and boil, when an odor of NH_3 proves **Nitrate.** If NH_4 salts be present we add a little KHO to the prepared solution, evaporate to dryness, heat the residue till no more fumes are evolved, and then dissolve in water and apply the zinc, etc., as above described.

Extra Step. Iodates being very difficult to detect in the preliminary, it is well to test for them specially (if they can possibly be present) by adding to the prepared solution KI and starch paste and acidulating with tartaric acid. This is not reliable in presence of nitrites.

Division D.—Course for the Detection of Organic Acids.

(*Only to be entered upon in the event of the original substance being proved to contain organic matter by charring on heating in the preliminary examination.*)

The solution to be used is that prepared for acidulous testing, as already described.

Step I. Place a minute fragment of litmus paper in a little of the prepared solution, and add acetic acid drop by drop with agitation until the paper *just* turns red, then take out the paper and add $AgNO_3$ in excess, lastly add a drop or two of very dilute NH_4HO till the precipitate *just commences* to redissolve. Now warm the tube in the Bunsen flame, when a reduction to metallic silver, forming a mirror on the tube=**Tartrate.**

Note.—The tube used must first be rendered chemically clean by boiling in it successively some dilute HNO_3 and then some dilute NaHO and rinsing with distilled water. **Formates** produce the same effects, but do not char on heating.

Step II. Place a minute fragment of test-paper into a portion of the prepared solution, and drop in dilute HCl till it *just* turns red, then dilute NH_4HO till it *just* turns blue again, cool thoroughly, add some $CaCl_2$ and shake well. If a precipitate forms (oxalate, tartrate, etc.) add excess of $CaCl_2$, shake, and let it stand in cold water for ten minutes and filter. Now add to the filtrate a little more NH_4HO and boil *gently* for some time, when a white precipitate=**Citrate.**

Note.—If $CaCl_2$ gives nothing in the cold of course we simply warm for the citrate straight off. As oxalate, tartrate, etc., have all been previously tested for, we shall know, before commencing to test for a citrate, whether we need to separate them in the cold or simply to add the NH_4HO and $CaCl_2$ and boil straight away.

The addition of rectified spirit to the solution in which boiling has failed to indicate citrate will bring down a *Malate* on cooling, but unless specially suspected this reaction is not a very certain one.

Step III. Place a fragment of test-paper into a portion of the prepared solution, and if alkaline, make it *exactly neutral* by carefully dropping in dilute HCl. Then apply the following tests to portions of this neutralised liquid :—

(*a*) Prepare some neutral ferric chloride, by adding very dilute NH_4HO to a solution of Fe_2Cl_6 until a permanent cloud just forms, and filtering.

Now add some of this reagent, and observe effect as follows :—

(1) Red color	Acetate Sulphocyanate Meconate Pyrogallate	(2) Purple color	Carbolate Sulpho-carbolate Salicylate
(3) Blue-black	Gallate Tannate	(4) Pinkish precipitate	Benzoate Succinate

Notes.

(1) Acetate, red, is instantly discharged by a drop of HCl; pyrogallate is turned black by excess of KHO and exposure to air. Sulphocyanate and meconate have been already proved in the inorganic acid course, but distinguished by action of $HgCl_2$ if desired.

(2) Acidulate a portion of prepared solution with HCl and shake up with ether. Remove the ether by a pipette and evaporate it on a watch-glass at a *very* gentle heat. Carbolic acid is left an oily liquid readily recognised, while salicylic acid is left in characteristic crystals, as giving a beautiful violet with Fe_2Cl_6. (Also see page 57 for another separation.) Sulphocarbolic acid gives no immediate precipitate with $BaCl_2$, but on evaporating with a little Na_2Co_3 and KNO_3 and fusing, then the residue dissolved in H_2O shows a sulphate with $BaCl_2$

(3) With excess of KHO a solution of gallic acid rapidly becomes dark on exposure to the air, while tannic acid gives a flocculent liquid not so rapidly changing. Tannic acid also precipitates solution of gelatine, and gallic does not.

(4) Take a good quantity of the neutralised and prepared solution, add excess of Fe_2Cl_6, filter out the precipitate and wash it. Now percolate it with some dilute NH_4HO, evaporate the liquid so obtained to a low bulk, cool thoroughly, and acidulate with HCl. *Benzoic acid* will separate in silky crystals, and succinic acid will not.

Step IV. If oleic, lactic, or sulphovinic (ethyl-sulphuric) acids be suspected, specially test for them as follows :—

Oleic acid will have shown its presence by always floating to the surface as an oily liquid whenever the prepared solution is acidulated with any acid. To confirm and distinguish it from the other fatty acids (stearic, etc.) take some of the prepared solution and acidulate with HCl, warm and set aside till the oily layer floats up. Now remove the liquid beneath, as far as possible, with a pipette, add some water, boil and drop in small fragments of K_2CO_3, until the oily layer is saponified and dissolves. Now put in a piece of test-paper and carefully add acetic acid to exact neutrality, then cool and precipitate, with excess of $Pb(C_2H_3O_2)_2$. Filter out the oleate of lead, wash it with boiling water, let it thoroughly drain, and then prove that it is soluble in ether (stearate and palmitate of lead are insoluble).

Lactic acid. Acidulate the prepared solution with HCl and shake up with ether. Pipette off the ether into a porcelain capsule and let it evaporate at a *gentle* heat, when the acid will be left and may be recognised as follows: (*a*) A portion heated burns at first with a blue flame, and then the flame becomes luminous as the temperature rises; (*b*) Another portion warmed with $K_2Mn_2O_8$ gives the odor of aldehyd.

Sulpho-vinates do not precipitate $BaCl_2$ in the cold, but, on boiling, give a precipitate of $BaSO_4$ and an odor of spirit.

SOLUBILITY.

An important consideration before proceeding to the Acidulous Course.

Looking to the metals found, the question arises, what acidulous radicals could possibly be there, supposing the original substance to be soluble or insoluble in water or acids, or totally insoluble. Of the solubility of the commoner salts, the following table will show this, and a careful study of it will save much unnecessary and often misleading testing for acids. **The usual radicals are in words and the rarer ones in symbols bracketed.**

Metals found.	If soluble in water, test for the following radicals.	If insoluble in water but soluble in acids, test for the following radicals.	If insoluble in acids, fuse with $KNaCO_3$, extract with H_2O, and test the solution for the following radicals.
Silver	Nitrate, Nitrite, Sulphate, Acetate (ClO, ClO_3, BrO_3, S_2O_3)	Oxide, Sulphide, Carbonate, Phosphate, Cyanide, Oxalate, Tartrate, Citrate (AsO_4, CrO_4, IO_3, S_2O_3)	Chloride, Iodide, Bromide
Mercury (ous)	Nitrate, Acetate, Sulphate (ClO, ClO_3, IO_3, BrO_3, NO_2, BO_3, $C_4H_4O_6$)	Oxide, Sulphide, Chloride, Iodide, Oxysulphate (PO_4, AsO_4, C_2O_4, CrO_4, $C_4H_4O_6$, $C_6H_5O_7$)	Chloride, Iodide, Bromide
Mercury (ic)	Chloride, Nitrate, Sulphate, Acetate (ClO, ClO_3, IO_3, BrO_3, Cy, S_2O_3, CrO_4, NO_2)	Oxide, Sulphide, Iodide, Carbonate, Oxysulphate (PO_4, C_2O_4, AsO_4, CrO_4, $C_4H_4O_6$)	Sulphide, Iodide
Lead	Acetate, Nitrate (Cl, I, ClO, ClO_3, BrO_3, NO_2, $C_6H_5O_7$)	Oxide, Sulphide, Carbonate, Phosphate, Oxalate (Cl, I, Cy, SO_3, S_2O_3, IO_3, BrO_3, AsO_4, CrO_4, $C_4H_4O_6$)	Sulphate, Chromate, Chloride, Iodide
Bismuth	Nitrate, Chloride, Sulphate, Acetate (Br, IO_3, BrO_3, ClO, ClO_3, S_2O_3, NO_2, $C_6H_5O_7$)	Oxynitrate, Oxychloride, Oxysulphate, Oxide, Sulphide, Carbonate, Phosphate (CrO_4, AsO_4, BO_3, C_2O_4, I, SO_4, $C_4H_4O_6$)	None
Copper (ic)	Chloride, Nitrate, Sulphate, Acetate (NO_2, ClO_3, S_2O_3, CrO_4, ClO, Br, IO_3, BrO_3, I, $C_4H_4O_6$, $C_6H_5O_7$)	Oxide, Sulphide, Carbonate, Phosphate, Arsenite, Oxyacetate (AsO_4, BO_3, Cy, SO_3)	"
Copper (ous)	Sulphate (Cl, Br)	Iodide (Cl, Br)	"
Cadmium	Chloride, Nitrate, Iodide, Sulphate (ClO, ClO_3, Br, IO_3, BrO_3, SO_3, S_2O_3, NO_2, $C_2H_3O_2$, $C_4H_4O_6$, $C_6H_5O_7$, BO_3)	Oxide, Sulphide, Carbonate, Phosphate (BO_3, C_2O_4, $C_4H_4O_6$)	"
Antimony	Chloride, Tartrate ($C_2H_3O_2$, C_2O_4)	Oxide, Sulphide, Oxychloride (SO_4, PO_4, CrO_4, I)	"
Tin (Stannous)	Chloride, Sulphate (ClO, ClO_3, Br, NO_2, NO_3, S_2O_3, $C_2H_3O_2$, $C_4H_4O_6$)	Oxide, Sulphide, Phosphate, Chromate (BO_3, C_2O_4)	"

Tin (Stannic)	Chloride (ClO_3, $C_2H_3O_2$, C_2O_4)	Oxide, Sulphide (SO_3)	None
Gold	Chloride	Sulphide (Cy)	,,
Platinum	Chloride	Sulphide	,,
Iron (Ferrous)	Chloride, Sulphate, Iodide (see Ferric, also SO_3, S_2O_3)	Oxide, Sulphide, Carbonate, Phosphate, Arseniate (C_2O_4, BO_3, CrO_4, Cy, $C_4H_4O_6$, AsO_3)	,,
Iron (Ferric)	Chloride, Nitrate, Sulphate, Acetate (Br, IO_3, BrO_3, ClO_3, NO_2, CrO_4, $C_4H_4O_6$)	Oxide, Sulphide, Iodide, Phosphate, Arseniate (C_2O_4 (when dried), BO_3, and SO_4 (and NO_3 when basic)	,,
Aluminium	Like Iron	Like Iron	,,
Cerium	Chloride	Oxide and Oxalate	Oxide
Chromium	Chloride, Sulphate, Acetate, Nitrate (Br, ClO, ClO_3, IO_3, BrO_3, S_2O_3, SO_3, C_2O_4, $C_4H_4O_6$, $C_6H_5O_7$)	Oxide, Phosphate, Arseniate (BO_3, CrO_4, Cy)	Oxide
Manganese	Chloride, Sulphate, Acetate, Nitrate (Br, ClO, ClO_3, IO_3, BrO_3, SO_3, S_2O_3, NO_2, CrO_4)	Oxide, Sulphide, Carbonate, Phosphate (AsO_4, C_2O_4, BO_3, Cy, $C_4H_4O_6$, $C_6H_5O_7$)	None
Zinc	Chloride, Sulphate, Acetate, Nitrate (ClO, ClO_3, Br, IO_3, BrO_3, NO_2, S_2O_3, SO_3, CrO_4, $C_6H_5O_7$)	Oxide, Sulphide, Carbonate, Phosphate (C_2O_4, BO_3, ASO_4, Cy, $C_4H_4O_6$)	,,
Nickel	Chloride, Nitrate, Sulphate, Acetate (ClO, ClO_3, IO_3, BrO_3, SO_3, S_2O_3, Br, NO_2)	Oxide, Sulphide, Carbonate, Phosphate (AsO_4, BO_3, CrO_4, Cy, $C_4H_4O_6$, $C_6H_5O_7$, C_2O_4)	,,
Cobalt	Chloride, Nitrate, Sulphate, Acetate (ClO, ClO_3, IO_3, BrO_3, SO_3, S_2O_3, Br, NO_2, $C_6H_5O_7$)	Oxide, Sulphide, Carbonate, Phosphate (AsO_4, BO_3, CrO_4, C_2O_4, Cy)	,,
Barium	Chloride, Nitrate, Acetate, Oxide (slightly) (ClO, ClO_3, I, Br, IO_3, BrO_3, S_2O_3, NO_2, Cy, $C_6H_5O_7$, S)	Carbonate, Phosphate, Oxalate, Chromate (O, AsO_4, BO_3, SO_3, $C_4H_4O_6$)	Sulphate
Strontium	Like Barium (except $C_6H_5O_7$)	Like Barium	Sulphate
Calcium	Like Barium (except $C_6H_5O_7$ soluble in cold but not in boiling water, $CaSO_4$ also slightly soluble)	Like Barium (except SO_3, which is soluble in H_2O and $CaSO_4$ slightly soluble)	Sulphate
Magnesium	Like Calcium	Like Calcium	None
Lithium	Oxide, Chloride, Sulphate	Carbonate, Phosphate, Oxalate	,,
Potassium	All radicals form soluble salts except those mentioned opposite	$PtCl_4 2KCl$, $KHC_4H_4O_6$, and K_2SiF_6	,,
Sodium	All radicals form soluble salts except	Sodium Antimoniate	,,
Ammonium	Like Potassium		

NOTE.—When a radical is mentioned in more than one column, it means that it is so slightly soluble as to sometimes appear insoluble at first sight.

§ V.—SPECIAL PROCESSES FOR PROVING THE IDENTITY OF CERTAIN READILY RECOGNISABLE SUBSTANCES.

(*A*) Salts of the Halogens.

Chlorine water. Characteristic odor; entirely volatile; bleaches indigo; KI+starch paste gives blue; $AgNO_3$ gives curdy white.

Hydrochloric acid. Slight fumes with sharp odor and strong acidity; entirely volatile; $AgNO_3$ gives curdy white; heated with MnO_2 evolves Cl_2.

Ferric chloride in solution. Orange-red liquid, not becoming milky with H_2O; test for ferric iron and for a chloride.

Antimonious chloride in solution. Orange-red liquid, becoming milky when diluted with H_2O; milky liquid divided; one part becomes orange with H_2S, and the other portion, after filtration, gives test for chloride with $AgNO_3$.

Mercurous chloride. Heavy dull white powder, entirely volatile; turns grey when boiled with $SnCl_2$, turns black when boiled with dilute KHO, and, after filtration, the liquid gives test for chloride on acidulating with HNO_3 and adding $AgNO_3$.

Mercuric-ammonium chloride. Opaque white powder, entirely volatile; becomes grey when boiled with $SnCl_2$; boiled with KHO turns yellow and gives off NH_3, and the liquid, after filtration, gives test for chloride on acidulating with HNO_3 and adding $AgNO_3$.

Chlorinated lime. Dull white powder, having the odor of ch orine; shaken up with water partly dissolves, and the solution, after filtration, gives with oxalic acid a white precipitate and evolves Cl_2.

Potassium chlorate. In crystalline plates, deflagrating on heating with evolution of oxygen and exploding when touched with H_2SO_4 with an odor of Cl_2O_4; solution gives no precipitate with $AgNO_3$; evaporated to dryness, heated to redness and the residue dissolved in very little water, yields the tests for potassium with $PtCl_4$ and for chloride with $AgNO_3$.

Mercurous iodide. Yellowish-green heavy powder, entirely volatile; heated gently in a dry tube gives sublimate of HgI_2, leaving a globule of metallic Hg.

Mercuric iodide. A bright scarlet powder turning yellow and then volatilising when gently heated on a piece of paper; boil with dilute KHO, let settle, and pour off, when the liquid gives blue with starch paste and HNO_3, while the insoluble portion dissolved in HCl and boiled with $SnCl_2$ gives grey.

(*B*) Oxides, Carbonates, etc.

Hydrogen peroxide. Colorless liquid with an odor similar to weak chlorine water; gives a blue with KI and starch paste, and the resulting liquid is *alkaline* to test-paper; no precipitate with $AgNO_3$.

Barium oxide or hydrate, and Calcium oxide or hydrate. Whitish powders which, when shaken up with a *little* water, do not *appear* to be soluble, but render the liquid strongly alkaline; add more water, shake and filter, and apply to the filtrate the tests for Ba and Ca, and in another portion prove soluble hydrate by getting a brown with $AgNO_3$; heated in a dry tube give off moisture if hydrates, but not if oxides.

Magnesium carbonate or oxide. Light white powders not changing color on heating; dissolve in dilute HCl (former effervesces, latter not), and then apply tests for Mg by adding $NH_4Cl+(NH_4)_2CO_3$ and getting no precipitate, and lastly adding Na_2HPO_4 and getting a white precipitate.

Zinc carbonate or oxide. White powders turning yellow while hot and white again on cooling; dissolve in dilute HNO_3 (effervescing or not), and apply tests for Zn.

Plumbic carbonate. Heavy white powder, turning yellow on heating; dissolved in smallest possible excess of HNO_3 effervesces, and the solution diluted with H_2O does not become milky; apply tests for Pb.

Plumbic oxide. Yellow or flesh-colored powder, heavy, and becoming yellow on heating; dissolves in HNO_3 without effervescence, and otherwise behaves as the carbonate.

Red lead. Heavy red powder turning yellow on heating, but *not* volatile; heated with dilute HNO_3 turns brown, and liquid poured off gives tests for Pb.

Plumbic peroxide. Puce brown powder, turning yellow when heated; heated with HCl evolves Cl_2, and the residue dissolved in boiling water gives yellow with K_2CrO_4.

Bismuth oxycarbonate and oxide. Former white powder, latter lemon-yellow, and both turning deep orange while hot and pale yellow on cooling; dissolve in smallest possible excess of HNO_3, when former slightly effervesces and latter not, and the solution diluted with H_2O becomes milky, proving Bi.

Ferric oxide. Reddish-brown powder unchanged by heat; dissolves in HCl without effervescence or odor, and the solution diluted gives the tests for Fe_2. *Hydrated* ferric oxide heated in a dry tube gives off moisture.

Ferroso-ferric oxide. Blackish-brown powder, turned lighter by heat; dissolves in HCl with effervescence or odor, and solution diluted gives tests for both Fe and Fe_2.

Potassium permanganate. Violet needles or prisms, giving off oxygen when heated and leaving a residue which when moistened is alkaline to test-paper; gives a fine violet solution turned colorless when warmed with HCl and spirit, the resulting liquid giving the tests for K and Mn.

Potassium bichromate or chromate. The former is in orange crystals, evolving oxygen when heated, and the latter is in yellow crystals unaltered by heat. Both give the tests for chromate with $AgNO_3$ and $Pb(C_2H_3O_2)_2$, and when heated with HCl and spirit both turn green and the resulting liquid gives the tests for Cr and K.

Chromic anhydride. In crimson needles, giving off oxygen when heated and leaving green Cr_2O_3. A solution mixed with dilute spirit gives off the odor of aldehyd, and forms a deposit of green Cr_2O_3.

Antimonious oxide. A greyish-white powder, readily fusible by heat; insoluble in HNO_3, but soluble in heated strong HCl; the solution diluted with H_2O becomes milky and then turns orange with H_2S.

Mercuric oxide. A yellow or red powder entirely volatile by heat; heated in a dry tube gives off oxygen and forms a sublimate of metallic Hg; dissolve in dilute HCl and apply tests for Hg.

Argentic oxide. An olive-brown powder, giving off oxygen and leaving metallic silver when heated; dissolve in dilute HNO_3 and apply the tests for Ag.

(*C*) Sulphates, etc.

Sulphur. A yellow or yellowish-white crystalline amorphous powder, which when heated takes fire, forms SO_2, and entirely burns away; insoluble in H_2O; boiled with HNO_3 dissolves with evolution of red fumes, and solution diluted gives white with $BaCl_2$.

Sulphuretted hydrogen water. Colorless solution with odor of H_2S and entirely volatile by heat; gives black with $Pb(C_2H_3O_2)_2$, etc.

Potassium polysulphide. A greenish deliquescent solid forming a greenish-yellow solution having the odor of H_2S. Acidulated with HCl and boiled gives off H_2S, deposits S_2, and the solution gives test for K with $PtCl_4$.

Kermes mineral. (Antimony sulphide, etc.) An orange-red powder, which when heated burns, giving off SO_2 and leaving a lighter-colored residue; entirely soluble in NaHO; soluble in boiling HCl with evolution of H_2S, and this solution diluted becomes milky and turns orange on adding H_2S; original powder boiled with solution of $KHC_4H_4O_6$ partly dissolves and the solution gives tests for Sb.

Sulphurous acid. Colorless liquid having odor of SO_2 and entirely volatile by heat; $BaCl_2$ gives no precipitate, or only a slight one, but on adding chlorine water a dense precipitate is produced by $BaCl_2$.

Sulphuric acid. A heavy, colorless, strongly acid liquid, which evaporated on a piece of paper chars it; entirely volatile by heat; $BaCl_2$ gives white precipitate; when diluted, dissolves metallic zinc with effervescence.

Mercuric sulphate. White crystalline heavy powder, entirely volatile by heat; treated with water forms a yellow insoluble powder, the liquid poured off from which gives white with $BaCl_2$; the yellow powder, dissolved in HCl, gives the tests for Hg.

(*D*) Borates, Nitrates, Phosphates, etc.

Boric acid. In minute tabular crystals, not volatile by heat, but fusible to a glassy residue which is acid to moistened test-paper; not appreciably soluble in cold water; soluble in rectified spirit and the solution burns with a green flame.

Nitric acid. A fuming liquid strongly acid and volatile by heat; warmed with metallic copper gives off red fumes.

Bismuth oxy-nitrate. Heavy white powder, turning orange on heating and becoming pale yellow on cooling; insoluble in water but soluble in HNO_3 and the solution diluted becomes milky; dissolved in equal parts H_2SO_4 and H_2O and a cold solution of $FeSO_4$ gently poured over the mixture, the characteristic nitrate ring is produced.

Calcium and sodium hypophosphites. White granular powders which on heating take fire, giving dense white fumes; if heated on platinum foil, go through it; dissolve and apply tests for Ca, Na, and hypophosphite.

Phosphoric acid. Colorless strongly acid liquid leaving on evaporation a non-volatile glassy residue, also acid to moistened test-paper; no precipitate with $AgNO_3$, but on carefully adding NH_4HO a yellow precipitate is formed.

Calcium phosphate. A light white powder unaltered by heat; soluble in dilute HNO_3, such solution giving a gelatinous precipitate with NaHO, insoluble in excess, but soluble in acetic acid; the solution so obtained, when divided, gives, in one part, a white with $(NH_4)_2C_2O_4$, and in the other a gelatinous white with Fe_2Cl_6.

Ferrous phosphate. A slate-blue powder turned reddish by heat; insoluble in H_2O but soluble in HCl; a portion of this solution, diluted and divided, gives the tests for iron, and the remainder mixed with tartaric acid and then with excess of NH_4HO gives a clear liquid which in turn yields a white precipitate with *magnesia mixture.*

Arsenious acid. A heavy white crystalline powder entirely volatile by heat, slightly soluble and feebly acid; $HCl + H_2S$ gives reaction for arsenic; $AgNO_3$ and $CuSO_4$ give no precipitates, but on carefully dropping in very dilute NH_4HO a yellow and a green precipitate are respectively produced.

Ferrous arseniate. A pale yellowish-green powder turned dark red by heat: dissolved in HCl gives tests for Fe; boiled with dilute KHO and filtered, the filtrate after *exact* neutralisation by dilute HNO_3 gives a red precipitate with $AgNO_3$.

(*E*) Organic Salt

Salt of sorrel (Potassium binoxalate). A white crystalline powder burning when heated, but not charring to any extent, and leaving an ash that is alkaline to moistened test-paper; the ash dissolved in HCl gives test for K with $PtCl_4$; the original dissolved in water by the aid of NH_4HO gives the test for oxalate with $CaCl_2$.

Cream of tartar (potassium bitartrate). A white gritty powder, charring by heat and leaving a black ash which is alkaline to moistened test-paper; the ash is tested by K as in the last case; the original heated with H_2SO_4 chars strongly and gives the odor of burning sugar.

Rochelle salt and *neutral potassium tartrate.* Both are freely soluble and char when heated, leaving an alkaline ash; a strong cold solution acidulated with acetic acid lets fall a crystalline precipitate of $KHC_4H_4O_6$, and the supernatant liquor gives with the former a yellow and with the latter a violet flame test.

Tartar emetic (potassium antimonyl tartrate). Chars on heating, leaving an alkaline ash; solution acidulated with HCl gives a white cloud soluble in excess, and H_2S then gives an orange precipitate; the original gives the tests for a tartrate.

CHAPTER V.

QUALITATIVE DETECTION OF ALKALOIDS AND OF THE SO-CALLED "SCALE" MEDICINAL PREPARATIONS USUALLY CONTAINING THEM, WITH A GENERAL SKETCH OF TOXICOLOGICAL PROCEDURE.

DIVISION A. COURSE FOR THE DETECTION OF THE ALKALOIDS AND ALKALOID SALTS USED IN MEDICINE.

Note.—Aconitine and atropine are omitted because they can only be really detected by experiments upon animals, which are now illegal except by special licence. Salicine, although not an alkaloid, is included for convenience, as it may be mixed with quinine.

In this course not more than two definite *tests* for each alkaloid are recorded, and for the remaining tests the reader is referred to the full table in Division D of this chapter.

Step I. Heat on platinum foil. If the substance at once takes fire and burns away with a smoky flame and an odor of singed hair, it is probably an alkaloid.

Step II. Put a piece of red litmus paper on a watch-glass, lay on to this a little of the substance, and moisten it with a few drops of strong rectified spirit. If, on standing for a short time, the paper is rendered blue, we are dealing with a free alkaloid; if not, then it is an alkaloid salt, and in the latter case we shall have to search for the acid as well as the base.

Note.—Acetates of alkaloids often become basic and consequently alkaline by keeping, so beware of this.

Step III. To a fragment of the substance on a watch-glass (placed over white paper) add a drop of strong H_2SO_4, and stir:—a bright red=**Salicine** and a deep red=**Veratrine.**

Confirm this latter by getting a yellow with HNO_3 and a blood-red on warming with HCl.

Note.—In each step any colors other than those herein recorded are to be disregarded. Many alkaloids give pale dirty pinks with H_2SO_4.

Step IV. To the liquid in which H_2SO_4 has given no distinct red add a small fragment of powdered ammonium molybdate, and stir.

(*a*) Greenish purple } =**Morphine** or **Apomorphine.**
Greenish black }

Confirm by adding HNO_3 to another fragment, when pale red=morphine and fine purple=apomorphine. Further confirm morphine with Fe_2Cl_6 (blue) and with $H_2SO_4+Na_2HAsO_4$ (bluish-green).

(*b*) Bright orange-red=**Brucine.**
Confirm by testing another fragment with HNO_3 and getting a bright red, turned to violet on warming with $SnCl_2$.

(*c*) Bright greenish-blue=**Codeine.**
Confirm by adding HNO_3 to another portion=pale evanescent yellow.

(*d*) A yellowish-green=**Physostigmine.**
Confirm by adding HNO_3 to another portion=strong gamboge yellow. Further confirm by getting a red with KHO, becoming blue on evaporating to dryness on the water bath, and dissolving in HCl to a dichroic solution.

Step V. Treat another fragment with a drop of H_2SO_4 as before, then let another drop fall near it. Into the second drop put a fragment of powdered potassium bichromate, let it digest a moment and then stir the drops together.

(*a*) Beautiful violet (evanescent)=**Strychnine.**

(*b*) Emerald green after standing some time=**Caffeine** or **Pilocarpine.**
Test another portion for caffeine by adding a crystal of $KClO_3$ and a drop or two of HCl, evaporating to dryness and getting a red residue becoming purple with NH_4HO.

Note.—Do not decide too hurriedly about pilocarpine; set the glass aside for half an hour, and then if the emerald green is *quite distinct and no cinchona alkaloid is present* you may conclude that pilocarpine is really there.

(*c*) Dirty pale-yellowish pink=**Cocaine.**

Note.—This test is not very good, but a minute drop of a dilute solution placed upon the tongue will cause tingling and numbness, and a strong solution will give a precipitate with $(NH_4)_2CO_3$ *soluble* in excess. (Aconitine, which also tingles the tongue, is not precipitated by $(NH_4)_2CO_3$.)

Step VI. Dissolve some of the original in water (using a drop or two of acetic acid to help solution if necessary) then add chlorine water and a gradual excess of NH_4HO. A clear green solution= **Quinine** or **Quinidine.** A white precipitate=**Cinchonine** or **Cinchonidine.**

Note.—In presence of salicylate the ordinary tests for quinine fail, and in this case it is necessary to dissolve in dilute hydrochloric acid, shake up with ether to remove salicylic acid, and then draw off the watery solution from beneath the ether and test it for quinine.

(*a*) To distinguish between quinine and quinidine.
Dissolve in hot water with a drop of dilute H_2SO_4 and cool, and if any crystals separate out they are probably sulphate of quinine and are rejected by filtration. (The proper proportion of water and alkaloid salt to start with is 1 of salt in about 25 of water. If you are dealing with a *free* alkaloid you must dissolve it in hot water with the smallest possible amount of dilute H_2SO_4 and then cool.) Now add very carefully dilute NH_4HO until the solution is as nearly neutral as possible without producing a permanent precipitate, cool perfectly, and add a few drops of saturated solution of rochelle salts, and stir well or shake. This will precipitate **quinine.** If not, then add a little saturated solution (or a small crystal) of KI, and shake, which will precipitate **quinidine.**

Note.—If on adding the chlorine water and then the ammonia *gradually*, a *perfectly clear* green solution results, then we need not search for quinidine, but put it down as quinine. The presence of quinidine always causes a milkiness in the green.

(*b*) To distinguish between cinchonine and cinchonidine.

Dissolve in water, by aid of HCl if necessary, then cool and carefully make as nearly neutral as possible with *very dilute* NaHO (if necessary), and then add saturated solution of Rochelle salt and shake, when a precipitate=**cinchonidine**. If not that, then add NH_4HO : white precipitate=**cinchonine.**

Note.—The general principles to keep in mind as to the cinchona alkaloids are :—

(1) That *in a neutral solution* rochelle salt precipitates quinine and cinchonidine as tartrates, leaving the others in solution. (2) After filtration (if necessary) the addition of potassium iodide and a little spirit precipitates quinidine as iodide, leaving cinchonine and the amorphous alkaloids (quinoidine, etc.) in solution. (3) From this solution excess of ammonia precipitates both, and, on shaking with ether the amorphous alkaloid passes into the ether and the cinchonine remains as a precipitate. (4) The separation of quinine and cinchonidine may be roughly performed by shaking up with ether, in the presence of a *very slight* excess of NH_4HO, and then corking the tube and letting it stand in cold water for some hours, when cinchonidine, if present, deposits in crystals, and quinine remains in the ether. For this, the ether must be in very small quantity, just so as to form a distinctly visible layer.

Step VII. If we believe that we are dealing with an alkaloid salt we must now proceed to test for the acid. The acid radicals usually present in alkaloidal salts of commerce are, chloride, sulphate, acetate, phosphate, citrate, meconate, nitrate and salicylate.

The first step will be to dissolve a little of the alkaloid salt in very dilute HNO_3, and test (1) for Cl by $AgNO_3$; (2) for SO_4 by $BaCl_2$; (3) for PO_4 by excess of ammonium molybdate and HNO_3.

The next will be to dissolve in water only, and test with Fe_2Cl_6 for acetate or meconate (red) or salicylate (violet). (Acetate decolorised by boiling, meconate not so; also acetate gives no precipitate with $Pb(C_2H_3O_2)_2$, and meconate does.)

Lastly, we must test in the usual way for a citrate (but unless the base be caffeine, this is not likely), and also for a nitrate (especially with pilocarpine and strychnine).

DIVISION B. QUALITATIVE ANALYSIS OF SCALE PREPARATIONS.

These commonly contain

Metals	Organic bases	Acids
Ammonium	Quinine	Tartaric
Iron	Cinchonine	Citric
Potassium	Cinchonidine	Pyrophosphoric (or hypo-)
	Strychnine	Sulphuric.
	Beberine	

Step I. Heat a little to redness on platinum foil, and observe the following possible cases :—

(*a*) If it entirely burns away we suspect **Beberine sulphate**. Test the original solution of the scale for sulphate by $BaCl_2$, as usual; and for beberine with KHO, getting a yellowish-white precipitate entirely dissolved by agitating the liquid with twice its volume of ether. This ether separated and evaporated to dryness leaves a yellow resinous-looking residue entirely insoluble in dilute HCl. If beberine sulphate be thus proved, go no farther.

(*b*) An ash is left: (a) Put a small fragment of the ash upon a piece of red litmus paper, moisten it with a drop of water, and, if it turns the paper blue, suspect **potassium**; (b) Dissolve the remainder of the ash in nitric acid, dilute and test with excess of ammonium molybdate for **phosphoric** acid.

Note.—If K be suspected, prove it by igniting some more of the scale, extracting the ash with very little boiling water, filtering, cooling, and adding $PtCl_4$.

Step II. Make a weak solution of the scale, acidulate it with a drop of HCl, and test for **ferrous** iron with $K_6Fe_2Cy_{12}$, and for **ferric** with K_4FeCy_6. Also test another solution of the scale by adding excess of $AgNO_3$, when a copious precipitate may form. Now add a drop of very dilute NH_4HO till the precipitate *just commences to dissolve* and heat, when reduction to black or a mirror = **Tartrate.**

Step III. Make a fairly strong solution of the scale, add excess of NaHO, boil, and smell for **ammonium.** If neither phosphoric nor tartaric acid has been already found, filter out the precipitated ferric hydrate and use the filtrate for testing for acids as follows:—

(*a*) Test a portion for citric acid exactly as directed in the organic acid course (page 80).

(*b*) Test another portion by exactly neutralising with HNO_3 and adding $AgNO_3$, when a white precipitate = **pyrophosphate,** and the same turning black = **hypophosphite.**

Of course this step is never to be taken unless an indication of P be got in the ash with molybdate in Step I.

Step IV. Make a solution of a fair amount of the scale, add a drop or two of very dilute NH_4HO, and then add some strong NH_4HO. Scales with strychnine only will not show a precipitate; with quinine they will give a precipitate with the dilute NH_4HO and this precipitate will dissolve in the strong; with cinchonine or cinchonidine a precipitate (white) will remain even with strong NH_4HO.

Case (*A*). There is either no precipitate, or it dissolves in strong NH_4HO:—Add some chloroform and shake up. Separate the chloroform by a pipette, and evaporate it to dryness in two portions on separate white dishes. Test the one residue for **strychnine** with H_2SO_4 and $K_2Cr_2O_7$, and the other for **quinine** with chlorine water and NH_4HO.

Case (*B*). The NH_4HO causes a permanent white precipitate:—Filter out precipitate, and dissolve it off the filter with a little warm water containing a few drops of acetic acid. Boil the solution down to a low bulk, cool, neutralise if necessary with very dilute NaHO, and then add a few drops of saturated solution of rochelle salts and shake, when a white precipitate = **cinchonidine.** If not that, then add NH_4HO, when a white precipitate insoluble on shaking with ether = **cinchonine.**

DIVISION C. GENERAL SKETCH OF THE METHOD OF TESTING FOR POISONS IN MIXTURES.

This course is only carried down to the best method of preliminary procedure for the *isolation* of the poison, all the individual tests to be afterwards applied having been already fully described in this or former chapters.

Step I. If the liquid be very strong, acid effervesces violently with $NaHCO_3$, test for poisonous acids, specially for **Nitric** and **Oxalic.**

Step II. Acidulate with $\frac{1}{7}$ of its bulk of HCl (filter if necessary), and apply Reinch's test for As, Sb, and Hg.

Step III. Burn to ash, dissolve this in HCl and test by ordinary course for poisonous metals, especially Pb, Cu, and Zn.

Step IV. If the original, either alone or when heated with dilute H_2SO_4, gives the odor of HCN or of carbolic acid, test specially for them.

Step V. *If the original has no odor of opium* we proceed to apply Stas's process for the detection of alkaloids as follows:—

If the original be a solid, it is operated upon directly, but, if a fluid, it is first evaporated to dryness on a water bath. Add some strong alcohol and a small crystal of tartaric acid, boil and filter. Evaporate the filtrate to dryness on the water bath, and take up with warm water slightly acidified with acetic acid, then cool and filter (if necessary), taking care that the liquid just remains acid. Now put this acid liquid into a separator (fig. 17), and shake it up with ether or benzene, and carefully separate the ether. (This ether may contain fat, certain bitter principles, and glucosides, and therefore, in a general investigation of a drug, it should not be rejected but evaporated, and the residue examined.) Now make the liquid distinctly alkaline by the careful addition of Na_2CO_3 or NaHO, and again shake up in the separator with chloroform, which will take up all the alkaloids except morphine The chloroform is separated, evaporated at a very gentle heat, and the residue tested for alkaloids by the course given in Division A or by the table given in Division D of this chapter. Lastly, the alkaline liquid is shaken up with amylic alcohol, which extracts morphine and leaves it upon evaporation.

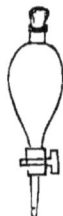

Note.—It is often better to get the alkaloids out from chloroform or amylic alcohol by shaking the separated solvent up with water acidulated with acetic acid or HCl, thus getting an aqueous solution and leaving any resinous matters in the chloroform.

Step VI. *When opium is suspected.* Acidulate with acetic acid and filter, if necessary (any alcohol present being got rid of by boiling it off). Precipitate when cold with solution of plumbic acetate, filter and preserve the precipitate (A) for examination for meconic acid and the filtrate (B) for morphine.

(*a*) Precipitate (A) is suspended in water and treated with H_2S till perfectly decomposed; the PbS filtered out, and the filtrate, after evaporation to drive off H_2S, tested for **meconic** acid. If this be found, it is held to be sufficient proof of presence of opium taken in connection with the odor of the original.

(*b*) Filtrate freed from Pb by H_2S and filtering is evaporated to dryness with a slight excess of $NaHCO_3$ on a water bath. The residue will yield its **morphine** to alcohol, generally in a state sufficiently pure to evaporate a drop, and test. If not, then amylic alcohol must be used.

DIVISION D. GENERAL RÉSUMÉ OF THE TESTS FOR ALL THE CHIEF ALKALOIDS.

The following tables are those given by Dragendorff in his "Pflanzenanalyse." The author has carefully repeated these tests, and has found them to be fairly accurately described, except aconitine, for which the reactions given are not characteristic. Reference to such a list will rarely be necessary, the working by Division A already given being amply sufficient for all ordinary purposes.

The following are the reagents used :—

1. Pure strong sulphuric acid free from nitrous fumes.
2. 200 parts of sulphuric acid with 1 part of nitric acid.
3. ·1 gramme of sodium molybdate in 10 c.c. strong sulphuric acid (Fröhde's test).
4. 1 part alkaloid mixed with 5 parts powdered white sugar, and then strong sulphuric acid dropped on.
5. Sulphuric acid and potassium bichromate used as already described in Division A.
6. Nitric acid (strong 1.3 sp. gr.)
7. Strongest fuming hydrochloric acid.
8. Ordinary solution as neutral as possible.

The reagents being generally added in turn to a fragment of the dry alkaloid, or to the residue left on evaporation of an alkaloidal solution.

TABLE OF GENERAL REACTIONS OF THE ALKALOIDS.

Alkaloid.	Pure H_2SO_4.	H_2SO_4 + HNO_3.	H_2SO_4 + sugar.	Fröhde's R.	HNO_3.	Conc. HCl.	H_2SO_4 + $K_2Cr_2O_7$.	Fe_2Cl_6.	Other Reactions.
Aconitine.. ..	grad. violet	grad. violet	fine red	yellow-brown	reddish-brown	no effect	..	in. aq. sol. yellow ppt.	A drop of a dilute solution causes tingling and numbness on tongue.
Atropine	no effect	no effect	no effect	no effect	no effect	no effect	discolored	not pptd.	Dilates the pupil.
Beberine	olive-green	olive-green	..	brown-green	brown-red				
Berberine ..	olive-green	olive-green	..	brown-green	brown-red			..	The addition of a few drops of chlorine water to a solution of berberine in excess of hydrochloric acid produces a red coloration.
Brucine .. .	no effect	red	no effect	red	red	no effect	orange		A solution in dilute sulphuric acid (1·8) is turned fine red by bichromate of potassium.
Caffeine	no effect	no effect	no effect	no effect	no effect	no effect	no effect	not pptd.	The residue obtained by dissolving in chlorine water and evaporating is colored red by ammonia.
Calabarine ..		..	..	..	..	..	..	..	
Chelidonine ..	no effect	green	..	..	..		..	..	The precipitate with potassio-mercuric iodide is insoluble in alcohol.
Cinchonine ..	no effect	no effect	no effect	no effect	no effect	no effect	..	..	Remain colourless when treated as above.
Cinchonidine ..	no effect	no effect	no effect	..	no effect	no effect	..	not pptd.	
Codeine	no effect	grad. blue	red	grad. deep blue	yellow	..	olive-green	no effect	Warmed with sulphuric acid and ferric chloride turns blue.
Colchicine ..	yellow	blue	yellow	yellow	blue	no effect	green, then brown	..	If the blue color produced by adding KNO_3 to the solution in H_2S_4 is allowed to fade, the residue is colored red by potash.
Colchiceïne ..	yellow	blue	yellow	yellow	blue	no effect	green, then brown		
Coniine	no effect	no effect	no effect	no effect	no effect	no effect	..		The hydrochloric acid solution leaves a crystalline residue when evaporated.
Curarine	red	red	red		..		blue, then violet and red		
Delphinine ..	no effect	no effect	no effect	no effect	remain light	remain light	..	..	Bromine colors the solution in sulphuric acid violet.
Delphinoïdine	red	no effect	green	blood-red	remain light	remain light	..	..	
Emetine	brown-green	brown-green	..	red and green	..	..	brown	..	Conc. hydrochloric acid colors the solution in Fröhde's reagent deep blue.
Gelsemine ..	yellow-red	..	red	..	..	..	..	..	Sulphuric acid and oxide of cerium color cherry-red.
Hyoscyamine..	no effect	no effect	no effect	no effect	no effect	no effect	discolored	not pptd.	Dilates the pupil.
Jervine	yellow, then light green	yellow, then light green	..	..	no effect	no effect			
Morphine.. ..	no effect	violet	red	violet	yellow	no effect	brown	blue	Reduces iodates, liberates iodine from iodic acid.
Narceïne	grad. grey, turning blood-red	yellow, turning orange	..	brown-green, red and blue	yellow	no effect	..	..	The crystals turn blue when moistened with dilute solution of iodine.
Narcotine ..	grad. raspberry	reddish-violet	--	..	..	no effect			Fine red is developed when the sol. in dil. sulph. acid is evaporated.
Nicotine	no effect	no effect	no effect	no effect	no effect	no effect			The hydrochloric acid solution leaves on evaporation an amorphous residue.
Nepaline	grad. red	grad. red	grad. red		..	..			
Papaverine ..	no effect	no effect	..		orange	no effect			Turns blue, warmed with sulphuric acid.
Physostigmine	grad. red	grad. red	..		red	reddish			Solution of chloride of lime colors red. The precipitate with potassio-mercuric iodide is soluble in alcohol. Physostigmine produces tetanus.
Pilocarpine ..	no effect	no effect	..	..	..	..	green		
Piperine ..	grad. green	grad. green	..	yellow, then brown	orange	..	..	ppt. in HCl. sol.	
Quinamine ..	..	..	..	..	..	..	..	..	The ppt. with chloride of gold is rapidly decomposed with red coloration.
Quinidine ..	no effect	no effect	no effect	greenish	no effect	no effect	..	..	Like quinine.
Quinine	no effect	no effect	no effect	greenish	no effect	no effect	light yellow	not pptd.	The chlorine-water solution is colored green by ammonia, red-brown with ammonia and ferricyanide of potassium.
Sabadilline ..	grad. fine cherry-red	grad. fine cherry red	grad. reddish-violet	grad. reddish violet	yellow	wine-red			Much weaker in physiological action than veratrine.
Sabatrine.. .	grad. fine cherry-red	grad. fine cherry-red	grad. reddish-violet	grad. reddish violet	yellow	wine-red			
Solanine	light reddish	light reddish	..	..	margin blue				A hot mixture of equal volumes of alcohol and sulphuric acid is colored red by solanine.
Staphysagrine	no effect	no effect	brown	violet-brown	..	no effect			
Strychnine ..	no effect	no effect	no effect	no effect	no effect	no effect	blue, passing quickly to violet and red	brownish-green ppt.	The blue coloration produced by H_2SO_4 and oxide of cerium is finer than with bichromate.
Taxine	red	..			..		..		The double chloride of taxine and gold is easily soluble.
Thalictrine ..	..	deep green							
Theobromine ..	no effect	no effect	no effect	no effect	no effect	no effect	no effect	not pptd.	Like caffeine.
Thebaine.. .	blood-red	red	..	orange	yellow	yellow			
Veratrine.. ..	grad. fine cherry-red	grad. fine cherry-red	grad. blue	grad. cherry-red	yellow	red	reddish-brown	ppt. in HCl. sol.	
Veratroidine ..	grad. fine cherry-red	grad. fine cherry.red	grad. violet	grad. cherry-red	yellow	red			

PART II.

QUANTITATIVE ANALYSIS.

CHAPTER VI.

WEIGHING, MEASURING, AND SPECIFIC GRAVITY.

I. WEIGHING AND MEASURING.

ALL bodies mutually attract each other. As the earth is the largest body within our atmosphere, it follows that its attraction is always greater than that of any surrounding matter. The force thus developed is called the attraction of gravitation, and its exercise is the cause of weight. Weighing is performed by means of the well-known appliance called the **balance.** Figure 18 illustrates a chemical balance of the modern short-beam type. H is the handle by which the balance is put into action, and R is the appliance for

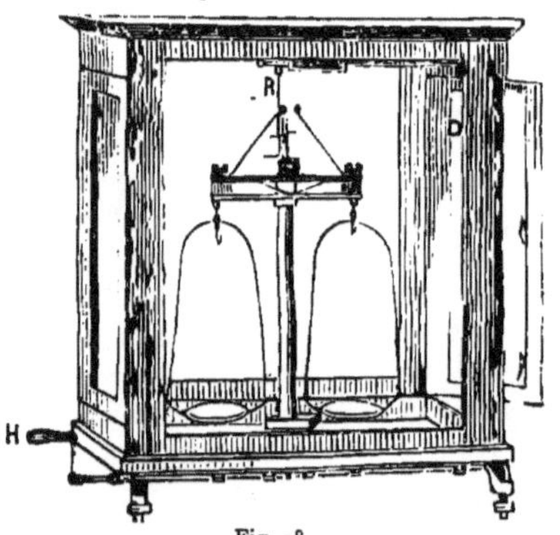

Fig. 18.

placing rider weights upon the graduated beam, aided by weights made either according to the English or the metrical system, as follows:—

(*a*) *The English system.*—In weights of precision, any amount above 10 grains is usually represented by a series of small brass cylinders, from 10 to 1000 grains; then follow 6, 3, 3, 2 and 1 grains in platinum wire, and afterwards ·6, ·3, ·3, ·2, and ·1 of a grain in platinum, or, more frequently, in aluminium wire. Quantities of less than $\frac{1}{10}$ grain are weighed by a small rider of gold wire placed on the beam of the balance. The foundation of the English system is the inch. One cubic inch of distilled water, measured at 60° F. and 30 inches barometrical pressure, weighs 252·45 grains, or 252½ grains nearly. There are 437·5 grains in an ounce, and 16 ounces (or 7000 grains) in a pound. Measure of capacity is obtained by weighing out 10 lb. of water at 60° Fahr. and 30 inches bar., when the whole measures one gallon. The gallon is in turn divided into 8 pints (=20 ounces, or 8750 grains of water, per pint); the pint into 20 fluid ounces (=437·5 grains of water per fluid ounce); the fluid ounce is divided into 8 fluid drachms (=54·68 grains of water per fluid drachm); and, lastly, the fluid drachm is divided into 60 minims (=·91 grain of water in each minim).

(*b*) *The Metrical system.*—The metrical weights of precision above one gramme are in brass; and then we have ·5, ·2, ·1, ·1, and following them ·05, ·02, ·01, ·01, all in platinum or aluminium foil. The quantities below ·01 (one centigramme) are weighed by a rider on the beam. The combination of 5, 2, 1 and 1 has been chosen because they have been found to give the greatest number of possible combinations with the fewest weights. Figure 19 shows such a box of metrical weights as usually employed in quantitative analysis. The metrical system is founded upon the *mètre.* The metre is multiplied and divided entirely by 10, thus:—

Fig. 19.

Kilo-metre	1000·
Hecto-metre	100·
Deca-metre	10·
Metre	1·
Deci-metre	·1
Centi-metre	·01
Milli-metre	·001

The metre taking the practical place of the English yard, the decimetre consequently takes the place of the foot, and the centimetre of the inch; and just as weight is got in our system from the cubic inch, so it is got metrically from the cubic centimetre, only much more simply, because 1 *cubic centimetre of distilled water, measured at* 4° C. *and* 760 *millimetres bar., weighs one gramme.* The gramme is multiplied and divided exactly as the metre, thus:—

Kilo-gramme	1000.
Hecto-gramme	100.
Deca-gramme	10.
Gramme	1.
Deci-gramme	.1
Centi-gramme	.01
Milli-gramme	.001

One kilogramme (1000 grammes) of water at the standard temperature and pressure measures one litre (or 1000 cubic centimetres), and we have therefore the following simple relation of weights and measures of water.

Weight.	Measure.
1000 grammes	1 litre or 1000 cubic centimetres
100 ,,	1 deci-litre or 100 ,, ,,
10 ,,	1 centi-litre or 10 ,, ,,
1 gramme	1 milli-litre or 1 cubic centimetre.

So we see that using water at 4°C., a gramme by weight and a cubic centimetre by measure amount to the same thing; as likewise do a kilogramme by weight and a cubic decimetre (or litre) by measure. The relation between the two systems is easily calculated from the following standards:—

Metrical.		English.
1 Gramme	=	15·432 grains.
1 Kilogramme	=	2·205 lb. (or 15,432 grains).
1 Litre	=	1·76 pints (or 35 fl. oz. 2 drachms 11 minims).
1 Metre	=	39·37 inches.

So that 1 decimetre is, as nearly as possible, 4 inches; and 1 decilitre, a trifle over 3½ fluid ounces.

II. SPECIFIC GRAVITY

may be generally explained to be the weight of anything as compared with that of an equal volume of something else taken as a standard. For liquids and solids the standard is distilled water at a temperature of 60°. An acquaintance with the various cases which may occur in the taking of specific gravity is of great importance, as it forms an exceedingly ready method of testing the purity and strength of many substances. A knowledge of the specific gravity of the various bodies also enables the chemist to tell at once what any given volume of a liquid ought to weigh, or conversely, what size of a vessel will be required to contain any given weight. The following are the chief varieties of cases which may occur in taking:—

(*A*) Specific Gravity of Liquids.

CASE 1. **To take the specific gravity of a fluid.**—A small bottle of thin glass is procured, and counterpoised upon a balance. It is then filled with distilled water at 60° F., and the weight of the water thus introduced noted. The bottle, having been emptied and dried, is filled with the liquid to be tested, also at 60° F., and the whole is again weighed. By this means, having ascertained the weight of equal bulks of water and fluid, it only remains to divide the weight of the fluid by the weight of the water, and the quotient will be the specific gravity required. To make the calculation clear, observe the following examples:—

A counterpoised bottle filled with distilled water weighs 1000 grains; the same bottle filled with sulphuric acid weighs 1843 grains, then:—

$$\frac{1843}{1000} = 1{\cdot}843, \text{ the specific gravity of the acid.}$$

Again, the same bottle, carefully washed, and filled with rectified spirit, weighs 838 grains, then:—

$$\frac{838}{1000} = {\cdot}838, \text{ the specific gravity of rectified spirit.}$$

In practice, bottles are sold with perforated stoppers, which, when entirely filled with the liquid, and the stopper dropped in, so that no bubbles of air are allowed to remain between the stopper and the liquid, exactly hold a given weight of water. A counterpoising weight for the empty bottle is also provided; so that there is nothing further to be done but simply to place the counterpoise in one scale, and the bottle, filled with the liquid under examination, in the other; and having ascertained the weight, to divide by the known weight of water for which the bottle was constructed. Fig 20 shows a specific gravity bottle in its case, with the counterpoise in a little

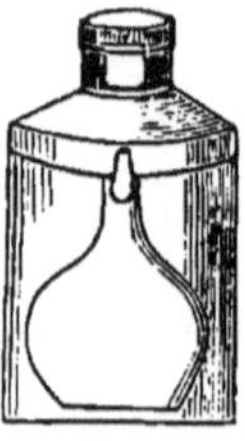

Fig. 20.

Fig. 21.

box on the lid. Fig. 21 shows a specific gravity bottle the stopper of which is a thermometer, thus enabling us to observe the exact temperature of the liquid at the moment of weighing.

CASE 2. **To take the specific gravity of a liquid by means of the hydrometer.**—The hydrometer depends for its action on the theorem of Archimedes. If a solid body be immersed in a liquid specifically heavier than itself, it continues to sink until it has displaced a bulk of fluid equal to its own weight, and then it becomes stationary. Suppose an elongated body with a weight at its base to cause it to float upright, which has a specific weight exactly half that of water, be immersed in that fluid, it will sink to exactly half its length, because its whole weight is counterpoised by a bulk of fluid equal to half its size. Hydrometers are long narrow glass or metal tubes with a bulb near the bottom filled with air, and another smaller bulb beneath containing a sufficient quantity of mercury to weight it and cause it to float upright. There are two kinds of hydrometers: (1) for fluids heavier than water, and (2) for fluids lighter than water. The graduation of the former is performed by immersing the instrument in water and introducing such a quantity of mercury as will cause it to sink, so that only about one inch remains unsubmerged, and marking this point 1. The instrument is then plunged successively into several liquids heavier than water, the specific gravities of which are known, and the points to which it rises are marked and numbered. By this means a scale can be made between those points, indicating any gravity from 1 upwards. In hydrometers for fluids lighter than water, the first sinking in that quid is continued by weighting until only the upper bulb is immersed; and this point having been marked 1, the instrument is placed successively in known fluids lighter than water, the points to which it sinks marked, and by this means a whole scale is obtained. The method of using the hydrometer is readily seen from the illustration (fig. 22), in which (A) is the hydrometer, and (B) is a thermometer also placed in the liquid to show the temperature. Most hydrometers being made to indicate specific gravity at 60° F., it follows that the liquid must either be first brought to that temperature before using the instrument, or else the temperature employed must be noted, and a calculation made, based upon the coefficient of expansion of the liquid in question.

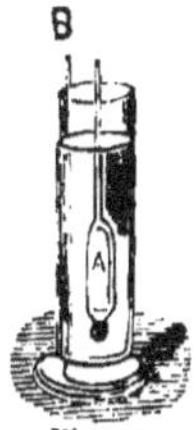

B

A

Fig. 22.

By Act of Parliament Sykes's hydrometer is used by the officers of excise to indicate the strength of spirituous liquors, and thus facilitate the collection of the revenue. It is a short brass instrument with the stem graduated from 0 to 10, and a series of nine weights to place beneath the bulb. By thus being able to change the weight, the length of the stem is only $\frac{1}{10}$ of that it would need to be with a permanent weight, and so the instrument will work in a proportionately smaller quantity of liquid. In using it, the thermometer must also be employed; and by observing (1) the temperature, (2) the weight put on, and (3) the point to which it sinks on the stem, and referring to a book of tables which is sold with the hydrometer, the strength of the spirit is ascertained.

Another modification of the instrument is found in Twaddell's hydrometer, which is used in this country for testing the density of liquids having a greater specific gravity than water. It is so graduated that the reading of any indicated degree, multiplied by 5 and added to 1000, gives the specific gravity as compared with water. Specific gravity beads form the only other variation of the hydrometric idea. These are small loaded bulbs of known specific gravities, which are thrown into the liquid to be tested, when the number marked upon the bead, which just floats underneath the surface and shows no tendency to sink or rise, gives the specific gravity required. Hydrometers in any form must in accuracy rank considerably beneath that of the specific gravity bottle; but in commercial operations, where an approximation only to correctness is required, these little instruments are invaluable.

Case 3. **To take the specific gravity of a liquid by weighing a solid body in it.**—Take the weight of a glass stopper, or other suitable plummet, by suspending it from the hook provided for the purpose in all balances of modern type (see fig. 12, page 6). Put a wooden stool (also provided with all modern balances) over the pan, and upon this place a beaker containing distilled water at 60° F. Let the plummet hang beneath the surface of the water and again weigh, and then empty out the water, substitute the fluid (also at 60° F.), immerse the plummet as before, and once more weigh. By deducting respectively the weights in water and in fluid from the weight in air, we get the loss of weight sustained by the plummet in each case. It is evident that the lighter the liquid, the more the plummet will weigh; therefore we *divide the loss of weight in the fluid by the loss of weight in water*, which will give the specific gravity of the liquid. This rule is now practically applied in all modern laboratories by means of the **Westphal balance** (fig. 23). By this a small thermometer (A), adjusted to a counterbalancing weight (B), is placed in the liquid, and the loss of weight is restored by little rider weights placed on the beam, which are so contrived as to readily indicate the specific gravity without calculation.

(*B*) Specific Gravity of Solids.

Case 1. **To take the specific gravity of a solid body in mass which is insoluble in and heavier than water.**—The method by which this process is conducted was suggested by a theorem attributed to Archimedes, which may be thus expressed:—A solid on being immersed in a liquid is buoyed up in proportion to the weight of the fluid which it displaces, and the weight it thus apparently loses is equal to that of its own bulk of the liquid. A piece of the solid substance to be tested is weighed, and is suspended by means of a fine thread from one arm of a balance so that it dips under the surface of a vessel containing distilled water at 60° F., when its weight is again noted. Its weight in water is deducted from its weight in air, and the weight in air is divided by the difference so obtained, which gives the specific gravity.

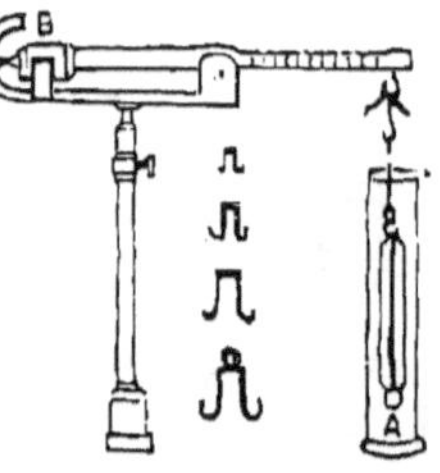

Fig. 23.

Example.

A piece of marble weighs . .	300 grains.
Immersed in distilled water . .	188·9 ,,
Difference in weight . . .	111·1 ,,

By dividing 300 by 111·1 we obtain the quotient 2·7, which is the specific gravity of the marble. The practical arrangement has been already described above (*Liquids, Case* 3).

Case 2. **To arrive at the specific gravity of a powder which is insoluble in and heavier than water.**—Weigh a portion of the powder in air, then introduce it into a counterpoised specific gravity bottle constructed to hold a known weight of water. Let the bottle be carefully filled with distilled water, gently agitating to insure that no minute bubbles of air shall remain attached to the particles of powder; then weigh the whole. From the weight of the powder in air, *plus* the known weight of water which the bottle should contain, deduct the weight obtained in the second operation, and divide the original weight of the powder by this difference.

EXAMPLE.—20 grains of a powder are weighed out, and poured into a counterpoised specific gravity bottle, constructed to hold 1000 grains of water. The bottle thus charged is found to weigh 1012 grains; then,

20 grains + 1000 grains	=	1020 grains.
Weight of the bottle when charged with powder and water . . .		1012 ,,
Difference		8 ,,

Therefore, 20 grains divided by 8 grains will give 2·5 as the specific gravity of the powder.

CASE 3. **To take the specific gravity of a substance in mass, insoluble in but lighter than water.**—The difficulty met with in this case consists in the impossibility of weighing such a substance alone in water, because it floats on the surface of that liquid. It therefore becomes necessary to attach a piece of lead sufficiently heavy to sink it, and thus a complication is introduced. The light substance is first weighed in air in the ordinary manner, and is then attached to a sinker, and suspended from one arm of a balance under the surface of distilled water, when the combined weight of both is ascertained. The light body is now detached, and the weight of the sinker alone in water noted. By these means we obtain the following data:—

i. The weight of the light body in air.
ii. The weight of the sinker in water.
iii. The weight conjointly of the light body and sinker in water.

We then deduct the weight of both in water from the weight of the sinker in water; add the weight of the light substance in air; and divide the weight of the light body in air by the product so obtained.

EXAMPLE.—A light substance weighs 120 grains in air; being attached to a piece of lead and weighed in distilled water the united weight amounts to 40 grains, while the weight of the lead alone in water shows 50 grains. Then:—

Weight of lead in water	50 grains.
Weight of both in water	40 ,,
Difference	10 ,,
Add weight of light body in air . .	120 ,,
Sum	130 ,,

Dividing 120, the weight in air, by 130 obtained as above, we arrive at the decimal fraction ·923 as the specific gravity of the light substance tested.

CASE 4. **To obtain the specific gravity of a substance soluble in water.**—Proceed exactly in the same manner as in Case 2 or 3, according as the body is in mass or in powder; but instead of water, use oil of turpentine or some other liquid in which the solid is insoluble. Having obtained the specific gravity of the substance by calculating just as if water had been used, multiply the result by the known specific gravity of the oil of turpentine or other fluid employed.

EXAMPLE.—A lump of sugar weighing 100 grains was found to weigh when immersed in oil of turpentine 45·62 grains. Then:—

The weight of the sugar in air was	100 grains.
,, ,, oil of turpentine . .	45·62
Difference	54·38

Dividing 100 grains by 54·38 grains yields 1·84 as the specific gravity as if water had been used; and by multiplying this result by ·87, the specific gravity of oil of turpentine, we obtain 1·6 as the actual specific gravity of the sample of sugar operated on.

Having thus considered in detail the various complications which may arise in taking the specific gravity of liquids and solids, it only remains to point out how the foregoing may be rendered subservient to commercial purposes.

(*C*) Practical Applications of Specific Gravity of Solids and Liquids.

CASE 1. **The specific gravity of a body being known, it is desired to ascertain the weight of any given volume of the substance.** Find the weight of the given bulk considered as water, and multiply this amount by the specific gravity.

EXAMPLE i.—What would be the weight of a fluid ounce of oil of vitriol? We know that a fluid ounce of distilled water weighs 437·5 grains, and the specific gravity of oil of vitriol is 1·843: so, if we multiply the former figures by the latter, we obtain 806·31 grains, which is the weight of a fluid ounce of this acid.

EXAMPLE ii.—How much should a litre of chloroform weigh? The weight of a litre of water is 1000 grammes; and by multiplying 1000 by 1·49, the specific gravity of the chloroform, we obtain 1490 grammes, as an answer to the question.

EXAMPLE iii.—How much should a fluid ounce of pure ether weigh? The specific gravity is ·72, and a fluid ounce of distilled water weighs 437·5 grains; multiplying the one number by the other gives 315 grains.

CASE 2. **Given the weight of any known bulk of a liquid, to find its specific gravity.** Divide the weight by that of the given bulk considered as distilled water.

EXAMPLE.—A pint of spirit weighs 16¾ ounces;—is it rectified or proof spirit? By dividing this weight by 20 ounces, the ascertained weight of a pint of distilled water, we obtain as an answer ·838. We know, therefore, that the spirit thus tested must have been rectified.

CASE 3. **To find the amount of solid matter, in grammes, present in 100 c.c. of a solution of given specific gravity.** So far as any ordinary rule can be laid down, especially with regard to saccharine liquids, for which this calculation is generally used, we multiply the gravity by 1000, and then, having deducted 1000 from the product, we divide by 3·95.

EXAMPLE.—A saccharine solution has a gravity of 1·0114: how much solid matter in grammes does it contain in each c.c.?

$$1{\cdot}0114 \times 1000 = 1011{\cdot}4 - 1000 = 11{\cdot}4 \therefore \frac{11{\cdot}4}{3{\cdot}95} = 2{\cdot}886 \text{ grammes per 100 c.c.}$$

(*D*) Specific Gravity of Gases.

Taking the density of gases and vapors involves many more complicated considerations than are required in the methods applicable to the specific gravity of liquids and solids. The standard adopted for such bodies is hydrogen, measured at a temperature of 0° C. and a barometrical pressure of 760 millimetres.

When taking the specific gravity of liquids or solids, it is easy to obtain the water or other fluid required at the exact temperature necessary, by the use of cooling or heating appliances. With a gas we need exercise no such manipulation, because the coefficient of expansion of all gases and vapors is alike and well ascertained. The measurement of gases is therefore conducted without any attempt to modify these conditions; but the indications of the thermometer and barometer being carefully noted at the time of the experiment, a simple series of calculations enables us to ascertain how much the volume of gas would have measured had the test been conducted at a standard of temperature and pressure. The following are specimens of such calculations:—

1. **Correction of the volume of gases for temperature.**—This calculation is based upon Charles's law, viz.: gases expand or contract one two-hundred-and-seventy-third ($\frac{1}{273}$) part of their volume for each degree of temperature,

Centigrade, through which their heat has been respectively raised or lowered. Therefore:—As 273 *plus* the temperature at the time of measurement is to 273 *plus* the required temperature, so is the volume of the gas at the period of measurement to the required volume.

For example:—The volume of a gas at 15° C. was 100 cubic centimetres: what would it be at the standard temperature of 0° C.? Then,—

$$\begin{array}{lcccc} \text{As } 273 & & 273 & & \\ \phantom{\text{As }}\underline{\ 15\ } & & \underline{\ \ 0\ \ } & & \\ \phantom{\text{As }}288 & : & 273 & :: & 100, \end{array}$$

which gives, as an answer, 94·795 cubic centimetres, the volume of the gas at standard temperature.

2. **Correction of the volume of gases for pressure.**—This calculation is based upon Boyle's law, viz.: gases expand or contract in volume in inverse proportion to the increase or diminution of the pressure,—that is to say, the greater the pressure the less the volume of gas, and the less the pressure the greater the volume of the gas.

For example:—At the moment of measuring 100 cubic centimetres of a gas, the barometer stood at 752 millimetres: what would the volume of the gas be at the standard pressure of 760 millimetres? Applying the rule of inverse proportion, we have:—As the required pressure is to the observed pressure, so is the observed volume to the required volume: then

$$760 \ : \ 752 \ :: \ 100,$$

which gives, as an answer, 98·95, the volume of the gas at standard pressure.

The manner in which the specific gravity of a **permanent gas** was formerly obtained was by exhausting a thin glass globe by means of the air pump and weighing it; then filling it with air at known temperature and pressure, and weighing; and lastly, pumping out the air, filling the globe with the gas at a similar temperature and pressure, and again weighing. After deducting the weight of the empty globe from each of the two latter weights, the weight of the gas was divided by that of the air.

Now, however, in modern laboratories all that is practically done away with, and the standard taken for the density of gases and vapors is hydrogen; because (1) it is the lightest known gas, and (2) we know the weight of any given volume of it without the necessity of weighing each time. Therefore to take the density of a gas or vapor we weigh a given number of cubic centimetres of the gas, noting the temperature and pressure at the moment of weighing, and having corrected the volume so obtained to 0° C. and 760 m.m. bar. we divide this by the weight of the same number of c.c. of hydrogen. A litre of hydrogen at 0° C. and 760 m.m. weighs ·0896 gramme bar.; therefore each c.c. of H will weigh ·0000896 gramme.

(*E*) Vapor Density.

After finding the percentage composition of organic bodies, and from that calculating an empirical formula (which is done by dividing the percentage of each element by its own atomic weight, then, taking the lowest of these answers as unity, dividing all the others by it and expressing the mutual ratios in the simplest full numbers) it is often necessary to prove whether the sum of such formula is the truemolecular weight. Upon the theory that all molecules occupy a space double that of an atom of hydrogen, we can prove our case by taking the density of a volume of the substance in vapor (if volatile) as regards hydrogen, and then this vapor density × 2 = the true molecular weight. This research acts as a check upon our formula obtained by analysis, and may or may not lead to our having to double it.

(*a*) **Dumas' process.**—A thin, clean, dry glass globe, about three inches in diameter, is employed. Its neck having been drawn out to a fine tube in the blowpipe flame, it is weighed, and the temperature and pressure noted. By gently heating the bulb and dipping the open end into the volatile liquid a suitable quantity is drawn into the globe by the contraction of the air. Attaching a handle by means of wire, the sphere is plunged into an oil bath furnished with a thermometer, and is then heated somewhat above its volatilising point. When all vapor has ceased to issue from the globe, the orifice is hermetically sealed, and the temperature and pressure again noted. The apparatus is allowed to cool, separated from the handle, cleansed, weighed, and the weight noted. The last step is to break off a fragment of the neck beneath the surface of a sufficiency of mercury, when, should the experiment have been carefully performed, the liquid enters the globe, completely filling it, and the capacity is ascertained by emptying its contents into a graduated glass measure. Supposing the experiment to have been perfectly successful, we have the following five data :—

1. Weight of globe filled with air.
2. Temperature and pressure at the time of weighing.
3. Weight of globe *plus* vapor.
4. Temperature and pressure at sealing.
5. Capacity of the globe.

Proceeding from these data, the first point is to find the actual weight of the globe. This is done by calculating the capacity of the globe from the temperature and pressure at the time of weighing to 0° C. and 760 mm. bar., and then multiplying the true volume thus found by ·001295, which is the weight of a cubic centimetre of air (1 litre at 0° C. and 760 m.m. bar.=1·295 gramme). Having thus obtained the weight of the air, it is deducted from the weight of globe and air, and the difference gives the true weight of the globe; and by deducting this latter from the weight of the globe *plus* vapor, we obtain the actual weight of the vapor. But as this weight is that of the volume of vapor to the temperature and pressure at the moment of sealing, it must be corrected to standard temperature and pressure, and the weight of an equal volume of hydrogen ascertained. To do this, the capacity of the globe is once more put down, and reduced from the temperature and pressure at sealing to 0° C. and 760 m.m. bar., and the resulting volume is multiplied by ·0000896, which is the weight of 1 cubic centimetre of hydrogen. The product, which gives the actual weight of an equivalent volume of hydrogen, is then taken, and divided into the weight of the vapor already found, and the answer is the density.

Fig. 24.

(*b*) **V. Meyer's Method.**—This is the simplest and most rapid process. The apparatus used is illustrated in fig. 24. The inner tube (A) is closed with a cork and arranged so that its bent delivery tube just dips under the surface of mercury contained in a trough. Any suitable liquid, which boils at a constant temperature, is placed in the outer tube (B), together with a thermometer, and heat being applied so as to boil the fluid, the air in the inner tube expands and passes off through the mercury. When bubbles of air cease to pass (showing that the air in the tube has been fully expanded to its proper volume at the given temperature), some water is poured upon the surface of the mercury, and a graduated "gas collecting tube" is filled with water and inverted over the delivery tube. A known weight of the substance is then introduced into the inner tube (A) by rapidly raising the cork, dropping the substance in, and instantly closing again. The vapor produced now displaces an equivalent volume of air, which passes into the measuring tube. When action ceases, the cork is opened to prevent back suction, and the air in the tube is measured, noting temperature and pressure. This volume in centimetres corrected to N.T.P., and multiplied by ·0000896, gives the weight of a volume of hydrogen equal to that of the vapor, and then by dividing the weight of the substance taken by such weight, we obtain the vapor density. The coefficient of expansion of all gases being equal, it is quite the same thing whether we measure an actual volume of vapor at a given temperature, or that of an equivalent volume of air displaced at the same temperature. Such a minute quantity of the substance must be taken as shall not, when in vapor, more than displace the air contained in the inner tube of the apparatus (which should hold about 100 c.c.), otherwise the whole process manifestly fails.

Methods of measuring gases at fixed temperature and pressure are given in Chapters IX. (ESTIMATION OF NITROGEN), and XII. (ANALYSIS OF GASES).

CHAPTER VII.

VOLUMETRIC QUANTITATIVE ANALYSIS.

I. INTRODUCTORY REMARKS.

VOLUMETRIC analysis is that in which the quantity of any *reagent* required to perform a given reaction is ascertained, and the amount of the substance acted upon is found by calculation. The process of adding the reagent from a graduated measure is called Titration.

(*A*) **Volumetric or standard solution** is a solution of definite strength made by dissolving a given **weight** of a reagent **in grammes** in a definite **volume** of water **in cubic centimetres** (or in grains and fluid grains). Such solutions are usually made by dissolving either a molecular weight of a reagent in grammes, or some decimal fraction of such weight, in 1000 c.c. (one litre) of water. The following abbreviations are used to express the strength of standard solutions :—

N = a normal solution	having	1	molecular	weight	in grammes	per	litre.
$\frac{N}{2}$ = a **semi-normal**	"	$\frac{1}{2}$	"	"	"	"	
$\frac{N}{10}$ = a deci-normal	"	$\frac{1}{10}$	"	"	"	"	
$\frac{N}{20}$ = a viginti-normal	"	$\frac{1}{20}$	"	"	"	"	

(*B*) **An indicator** is a substance added to enable us to ascertain, by a change of color (or other equally marked effect), the exact point at which a given reaction is complete.

The principal indicators employed are as follows :—

(*a*) *Solution of litmus*, which turns red with acids and blue with alkalies.

(*b*) *Alcoholic solution of phenol-phthalein*, which is colorless with acids but red with alkalies.

(*c*) *Starch mucilage*, which turns blue in presence of free iodine.

(*d*) *Solution of potassium chromate*, which gives a red with $AgNO_3$, but *not* until any halogen present has entirely combined with the silver.

(*e*) *Solution of potassium ferricyanide*, which ceases to give a blue color when any iron present has been fully raised to the ferric state.

(*C*) **General modus operandi.** A known weight of the substance to be analysed is accurately weighed, and having been dissolved or diluted with water (if necessary), the solution is placed in a flask, the indicator is added and the standard solution of the reagent is dropped in until the desired effect is attained.

The volume of the standard solution used is then noted; and its strength per 1000 c.c. being known, the actual amount of solid reagent that has been really added is easily found and calculated, by means of the equation for the action in question, to the amount of the substance under analysis it represents, as follows :—

Suppose, for example, we desire to ascertain the strength of a sample of caustic soda, and that we have weighed out 1 gramme, dissolved it in water, added litmus solution and found that it required 24 c.c. of standard solution of oxalic acid ($\frac{N}{2}$ = 63 grammes per 1000 c.c.) to just cause the color to change from blue to violet red (*i.e.* to neutralise it). Now, by the equation :—

$$\underbrace{H_2C_2O_4 \cdot 2H_2O}_{2)126} + \underbrace{2NaHO}_{2)80} = Na_2C_2O_4 + 4H_2O$$

63 — 40 = grammes of NaHO, equivalent to 1000 c.c. of oxalic acid solution (its strength being 63 grammes per 1000 c.c.).

Knowing this, we now ascertain how much NaHO is represented by the 24 c.c. of acid used; thus :—

$\frac{40 \times 24}{1000}$ = ·96 gramme of real NaHO present in the 1 gramme of caustic soda weighed out for analysis.

Then, if the results are to be expressed in percentage, we multiply the amount of the real article found by 100 and divide by the quantity weighed out for analysis, thus :—

$\frac{\cdot 96 \times 100}{1}$ = 96 per cent., strength.

Expressing the above calculations in rules to commit to memory, we have the following four steps :—

I. Write out the equation and reduce the first side of it to figures in molecular weights.

II. Cancel these weights down to equivalent weights corresponding to the indicated strength of the standard solution used (*i.e.* if $\frac{N}{2}$ divide by 2, if $\frac{N}{10}$ divide by 10, if $\frac{N}{20}$ divide by 20, etc.), thus obtaining the *equivalent* of the substance under analysis to 1000 c.c. of the standard solution.

III. Multiply this equivalent by the number of centimetres of standard solution used and divide by 1000.

IV. If percentage be required, multiply the last result by 100 and divide by the weight of substance taken for analysis.

Note.—It must be remembered that all waters of crystallisation must always be added to each substance containing them, in writing the equation for volumetric calculations.

(*D*) The apparatus specially employed in volumetric analysis.

1. *The measuring flask*, so constructed as to hold a definite amount of fluid (say 1000 or 100 c.c.) when filled up to the mark on the neck (fig. 25).
2. *The test mixer*, a cylindrical vessel, to hold 1 litre of fluid graduated in measures of 10 c.c. each (fig. 26).
3. *The burette*, a graduated tube, usually containing 100 c.c. and graduated in divisions of 1 c.c., for containing and delivering the standard solution. This is fitted with a clamp or stopcock at the bottom, which, when pressed or turned, allows the contained liquid to run out at any regulated speed desired. It should also be furnished with an appliance called "Erdmann's float," which enables us to read the quantity of fluid delivered

more accurately. (Fig. 27 shows two burettes in their stand as usually employed.)

4. *The pipette* is an instrument graduated to *deliver* a fixed volume of liquid (say 10, 20, 50 or 100 c.c.). Fig. 28 shows a set of such instruments arranged in a convenient stand.

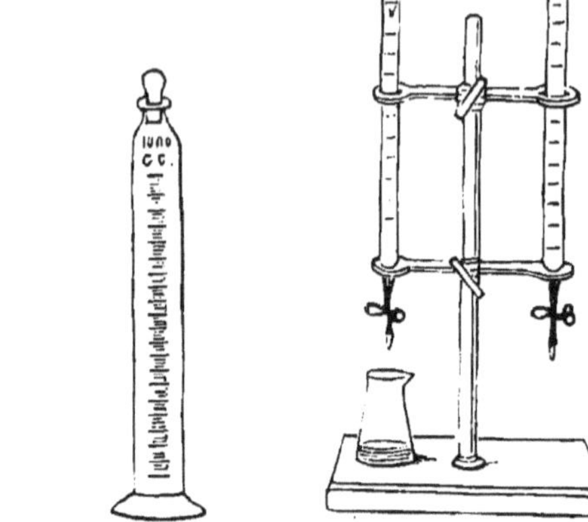

Fig. 25. Fig. 26. Fig. 27.

Fig. 28.

(*E*) **Weighing Operation**. The student should have a *tared* watch-glass for weighing out solids and a small stoppered bottle for weighing volatile liquids. By carefully keeping these much trouble is saved.

(1) *To weigh a solid.* Place the tared glass on the scale, and put on it what is judged to be a sufficient quantity of the article to be weighed, then weigh the whole and note the weight thus :—

Glass + substance	5·632	grammes.
Known tare of glass	5·132	,,
Weight taken for analysis	·500	,,

(2) *To weigh a volatile liquid.* Fill the small stoppered bottle with the liquid and weigh ; pour out what is judged to be sufficient into the flask containing the indicator and some water, replace the stopper and again weigh, noting each weight at the time thus :—

Total weight of bottle + fluid	20·982	grammes
Weight of bottle + fluid after pouring out .	15·482	,,
Weight of fluid taken	5·500	,,

Note.—It is most important always to take the weights directly down in a note-book from the balance, and to cultivate the habit of *always replacing the weights in their proper holes in the weight box when finished.* This enables us to have a double check, (1) from the weights in the pans, and (2) from looking at the empty holes in the weight box. In weighing, brass weights are used from 50 to 1 gramme ; flat platinum weights from ·5 to ·01 gramme, and the rider on the beam is used for milligrammes (*i.e.* ·009 to ·001). Before weighing, see that all the weights are in their right places in the box. At the conclusion of the weighing, read off the weights and put them down in a note-book, and then check that reading by putting them back into the box, looking, as you do so, at the note already made. Always close the case of the balance before using the rider, so as to prevent currents of air affecting the weight.

Having thus given a general idea of the mode of working, we now commence to practise with the chief standard solutions as follows.

II. STANDARD SOLUTION OF OXALIC ACID.

Strength $\frac{N}{2}$ = 63 *grammes per* 1000 *c.c.*

(*A*) Preparation.

This is made by powdering some *pure* oxalic acid, pressing it between the folds of blotting-paper (to remove any chance moisture), and weighing out exactly 63 grammes in a tared beaker. The powder is then washed out with distilled water from the beaker into the litre measuring flask, which is nearly filled with water and slightly warmed to aid solution. When all is dissolved, more water is poured in till the solution arrives at the mark in the neck of the flask, and finally the whole is cooled down to 60° F., and is once more exactly made up to the line with water.

This solution may then be used for the following purposes in the manner described under each case.

Check.—To check the strength of the standard oxalic acid itself. Take some *pure* $NaHCO_3$, and ignite it in a crucible for 15 minutes at a red heat, cool and weigh off 2·65 grammes of the resulting Na_2CO_3, and this should take exactly 50 c.c. of acid if truly $\frac{N}{2}$. The process is described below at (*C*).

(*B*) Estimation of Alkaline Hydrates.

By weighing out a definite quantity of the substance, diluting with, or dissolving in, water in a flask, adding a few drops of solution of litmus, and dropping in the standard acid from a burette until the last drop added just causes the blue to change to violet-red, the flask being agitated after each addition of the acid. In this way we should examine:—

(*a*) *Liquor ammoniæ fort. and liquor ammoniæ.*

Take about 3 grammes, and use the equation:—

$$\underbrace{H_2C_2O_4 \cdot 2H_2O}_{2)126 \over 63} + \underbrace{2NH_3}_{2)34 \over 17} \cdot H_2O = (NH_4)_2C_2O_4 + 4H_2O$$

17 grms. NH_3, equivalent to 1000 c.c. $\frac{N}{2}$ oxalic acid solution.

(*b*) *Potassium hydrate or sodium hydrate* or their *solutions.*

Take about 1 gramme of solid, or about 10 grammes of solution, and use the equations:—

$$\underbrace{H_2C_2O_4 \cdot 2H_2O}_{2)126 \over 63} + \underbrace{2NaHO}_{2)80 \over 40} = Na_2C_2O_4 + 4H_2O$$

40 = grms. real NaHO, equivalent to 1000 c.c. $\frac{N}{2}$ acid.

$$\underbrace{H_2C_2O_4 \cdot 2H_2O}_{2)126 \over 63} + \underbrace{2KHO}_{2)112 \over 56} = K_2C_2O_4 + 4H_2O$$

56 = grms. KHO, equivalent to 1000 c.c. $\frac{N}{2}$ acid.

(*c*) *Metallic sodium* (thrown on water and the resulting solution titrated).

Take under ·5 gramme, and use the equation:—

$$\underbrace{H_2C_2O_4 \cdot 2H_2O}_{2)126 \over 63} + \underbrace{Na_2}_{2)46 \over 23} = Na_2C_2O_4 + 2H_2O + H_2$$

23 grms. Na_2, equivalent to 1000 c.c. $\frac{N}{2}$ acid.

(*d*) *Lime-water and liq. calcis sacch.*

Take about 25 grammes, and use the equation :—

$$\underbrace{H_2C_2O_4 \cdot 2H_2O}_{\frac{2)126}{63}} + \underbrace{CaO}_{\frac{2)56}{28}} + H_2O = CaC_2O_4 + 4H_2O$$

28 = grammes CaO, equivalent to 1000 c.c. $\frac{N}{2}$ acid.

(*e*) *Borax.*

Take between 2 and 3 grammes, and use the equation :—

$$\underbrace{H_2C_2O_4 \cdot 2H_2O}_{\frac{2)126}{63}} + \underbrace{Na_2B_4O_7 \cdot 10H_2O}_{\frac{2)382}{191}} = Na_2C_2O_4 + H_2B_4O_7 + 12H_2O$$

191 = grammes borax, equivalent to 1000 c.c. $\frac{N}{2}$ acid.

(*C*) Estimation of Alkaline Carbonates.

By a similar process to (*A*), only conducted at a boiling temperature, so as to drive off all CO_2, and the standard acid to be added until two minutes' boiling fails to restore the blue of the litmus. Another and better method is to add a given volume of standard acid, more than sufficient to neutralise all the carbonate, then to boil until all carbon dioxide has passed off, and to titrate back with standard soda, using phenol-phthalein as indicator. The difference between the number of c.c. of acid taken and that of soda used will give the acid equivalent to the carbonate analysed.

(*a*) *Crystallised sodium carbonate.*

Take about 3 grammes, and use the equation :—

$$\underbrace{H_2C_2O_4 \cdot 2H_2O}_{\frac{2)126}{63}} + \underbrace{Na_2CO_3 \cdot 10H_2O}_{\frac{2)286}{143}} = Na_2C_2O_4 + CO_2 + 13H_2O$$

143 = grammes $Na_2CO_3 10H_2O$, equivalent to 1000 c.c. $\frac{N}{2}$ acid.

(*b*) *Dried sodium carbonate.*

Take about 1 gramme, and use the same equation less the 10 H_2O on each side :—

$$\frac{2)126}{63} + \frac{2)106}{53}$$

53 = grammes Na_2CO_3, equivalent to 1000 c.c. $\frac{N}{2}$ acid.

(*c*) *Sodium bicarbonate.*

Take about 3 grammes, and use the equation :—

$$\underbrace{H_2C_2O_4 \cdot 2H_2O}_{\frac{2)126}{63}} + \underbrace{2NaHCO_3}_{\frac{2)168}{84}} = Na_2C_2O_4 + 2CO_2 + 4H_2O$$

84 = grammes $NaHCO_3$, equivalent to 1000 c.c. $\frac{N}{2}$ acid.

(*d*) *Potassium carbonate* (commercially considered as always holding 16% of moisture).

Take about 2 grammes, and use the equation :—

$$\underbrace{H_2C_2O_4 \cdot 2H_2O}_{\frac{2)126}{63}} + \underbrace{K_2CO_3 \cdot 16\% H_2O}_{\frac{2)164 \cdot 28}{82 \cdot 14}} = K_2C_2O_4 + CO_2 + xH_2O$$

82·14 = grammes commercial K_2CO_3, equivalent to 1000 c.c. $\frac{N}{2}$ acid.

(*e*) *Potassium bicarbonate.*

Take about 2 grammes, and use the equation :—

$$\underbrace{H_2C_2O_4 . 2H_2O}_{\frac{2)126}{63}} + \underbrace{2KHCO_3}_{\frac{2)200}{100}} = K_2C_2O_4 + 2CO_2 + 4H_2O$$

100 = grms. $KHCO_3$, equivalent to 1000 c.c. $\frac{N}{2}$ acid.

(*f*) *Commercial "carbonate of ammonia."*

Take about 1 gramme, do not boil too violently, and use the equation :—

$$\underbrace{3(H_2C_2O_4 . 2H_2O)}_{\frac{6)378}{63}} + \underbrace{2(N_3H_{11}C_2O_5)}_{6)314} = 3(NH_4)_2C_2O_4 + 4CO_2 + 8H_2O$$

$52·3$ = grms., equivalent to 1000 c.c. $\frac{N}{2}$ acid.

(*D*) Estimation of Lead.

Weight out the substance, dissolve it in plenty of water (the flask $\frac{1}{3}$ full), with a drop or two of acetic acid to clarify it, and then carefully drop in the standard acid till precipitation ceases. Thus we operate upon :—

(*a*) *Plumbic acetate.*

Take 3 grammes, and use the following equation :—

$$\underbrace{H_2C_2O_4 . 2H_2O}_{\frac{2)126}{63}} + \underbrace{Pb(C_2H_3O_2)_2 . 3H_2O}_{2)379} = PbC_2O_4 + 2HC_2H_3O_2 + 5H_2O$$

$189·5$ = grms., equivalent to 1000 c.c. $\frac{N}{2}$ acid.

(*b*) *Liquor plumbi subacetatis.*

Take about 10 grammes, and use the equation :—

$$\underbrace{2(H_2C_2O_4 . 2H_2O)}_{\frac{4)252}{63}} + \underbrace{Pb_2O(C_2H_3O_2)}_{4)548} = 2PbC_2O_4 + 2HC_2H_3O_2 + 5H_2O$$

137 = grms., equivalent to 1000 c.c. $\frac{N}{2}$ acid.

(*E*) Estimation of Organic Salts of the Alkalies.

Organic salts of potassium or sodium are examined by weighing out about 2 grammes in a *tared* platinum or porcelain crucible, and then heating to redness in contact with the air until all is perfectly charred. The crucible is now cooled, and its contents dissolved in boiling water and filtered into a flask, and the filter washed with boiling water until the washings do not affect red litmus paper. The contents of the flask are then colored by litmus solution and titrated at boiling temperature with the standard acid, as already described in the case of alkaline carbonates. The ignition causes the conversion of the organic salt into an alkaline carbonate. Thus we should operate upon :—

(*a*) *Cream of tartar.*

$$\underbrace{2KHC_4H_4O_6}_{376} + 5O_2 = \underbrace{K_2CO_3}_{138} + 7CO_2 + 5H_2O ;$$

then $$\underbrace{K_2CO_3}_{138} + \underbrace{H_2C_2O_4 . 2H_2O}_{126} = K_2C_2O_4 + CO_2 + 3H_2O ;$$

therefore $$\underbrace{H_2C_2O_4 . 2H_2O}_{\frac{2)126}{63}} = \underbrace{2KHC_4H_4O_6}_{\frac{2)376}{188}}$$

63 = 188 = grms. of $KHC_4H_4O_6$, equivalent to 1000 cc. $\frac{N}{2}$ acid.

(*b*) *Neutral potassium tartrate.*

$$\underbrace{2K_2C_4H_4O_6 . H_2O}_{488} + 5O_2 = \underbrace{2K_2CO_3}_{276} + 6CO_2 + 4H_2O;$$

$$\text{then } \underbrace{2H_2C_2O_4 . 2H_2O}_{252} + \underbrace{2K_2CO_3}_{276} = 2K_2C_2O_4 + 2CO_2 + 4H_2O;$$

$$\text{therefore } \underbrace{2H_2C_2O_4 . 2H_2O}_{\substack{4)252 \\ 63}} = \underbrace{2K_2C_4H_4O_6 . H_2O}_{\substack{4)488 \\ 122}}$$

63 = 122 = grms. of $K_2C_4H_4O_6 . H_2O$, equivalent to 1000 cc. $\frac{N}{2}$ acid.

(*c*) *Rochelle salt.*

$$\underbrace{2(KNaC_4H_4O_6 . 4H_2O)}_{564} + 5O_2 = \underbrace{2KNaCO_3}_{244} + 6CO_2 + 8H_2O;$$

$$\text{then } \underbrace{2KNaCO_3}_{244} + \underbrace{2(H_2C_2O_4 . 2H_2O)}_{252} = 2KNaC_2O_4 + 2CO_2 + 4H_2O;$$

$$\text{therefore } \underbrace{2(H_2C_2O_4 . 2H_2O)}_{\substack{4)252 \\ 63}} = \underbrace{2KNaC_4H_4O_6}_{\substack{4)564 \\ 141}}$$

63 = 141 = grms. of $KNaC_4H_4O_6 . 4H_2O$, equivalent to 1000 c.c. $\frac{N}{2}$ acid,

(*d*) *Potassium citrate.*

$$\underbrace{2K_3C_6H_5O_7}_{612} + 9O_2 = \underbrace{3K_2CO_3}_{414} + 9CO_2 + 5H_2O;$$

$$\text{then } \underbrace{3K_2CO_3}_{414} + \underbrace{3(H_2C_2O_4 . 2H_2O)}_{378} = 3K_2C_2O_4 + 3CO_2 + 5H_2O;$$

$$\text{therefore } \underbrace{3(H_2C_2O_4 . 2H_2O)}_{\substack{6)378 \\ 63}} = \underbrace{2K_3C_6H_5O_7}_{\substack{6)612 \\ 102}}$$

63 = 102 = grms. of $K_3C_6H_5O_7$, equivalent to 1000 cc. $\frac{N}{2}$ acid.

(*F*) B.P. Standards of Strength by Oxalic Acid.

ARTICLE	B.P. STRENGTH
Liq. Ammon. fort.	32% NH_3.
Liq. Ammon.	10% NH_3.
Officia ammon. carb.	100% $N_3H_{11}C_2O_5$.
Borax	pure (100%).
Plumbic acetate	pure (100%).
Liq. plumbi subacetatis	about 24% $Pb_2O(C_2H_3O_2)_2$.
Lime-water	5 grains CaO in 10 fluid ounces.
Sodium (metallic)	95·7%.
Potassium and sodium hydrates	90 to 100%.
Potassium carbonate	97 to 100% of K_2CO_3+16% H_2O.
Crystallised sodium carbonate	96 to 100% of $Na_2CO_3 . 10 H_2O$.
Potassium and sodium bicarbonates	pure (100%).
Potassium tartrate and rochelle salt	99%.
Cream of tartar	92% $KHC_4H_4O_6$ in the salt dried at 212°; remainder only calcium tartrate.
Potassium citrate	pure (100%).
Liquor potassæ	5·84%.
Liquor sodæ	4·10%.

III. STANDARD SOLUTION OF SODIUM HYDRATE.

N=40 *grammes NaHO in* 1000 *c.c.*

(*A*) Preparation and Check.

As commercial soda is not pure, this solution has to be made as follows:—Dissolve 45 grammes of ordinary caustic soda in 1 litre of distilled water, and let it cool to 60° F. Now place 20 c.c. of standard solution of oxalic acid in a flask, add litmus, and run in some of this crude soda solution from a burette until neutrality is produced. Note the number of c.c. of soda used, put 50 times that volume of the crude solution into a test mixer, and make up to the 1000 c.c. mark with distilled water. On again checking, 20 c.c. of the acid should take exactly 20 c.c. of the perfected soda for exact neutralisation.

(*B*) Acidimetry.

Soda solution is used for taking the strength of acids by simply weighing out a quantity of the acid, and then running in the soda in presence of the litmus indicator if the acid be inorganic, or of phenol-phthalein if it be organic. We operate, as a rule, upon about 1 gramme of a strong or a solid acid, and from 5 to 10 grammes of a diluted acid. The following are some of the more important equations:—

(*a*) $\underbrace{NaHO}_{40} + \underbrace{HCl}_{36\cdot5} = NaCl + H_2O$

36·5 grms. of HCl, equivalent to 1000 c.c. NaHO.

(*b*) $\underbrace{NaHO}_{40} + \underbrace{HNO_3}_{63} = NaNO_3 + H_2O$

63 = grms. of HNO_3, equivalent to 1000 c.c. NaHO.

(*c*) $\underbrace{NaHO}_{40} + \underbrace{HC_2H_3O_2}_{60} = NaC_2H_3O_2 + H_2O$

60 = grms. of $HC_2H_3O_2$, equivalent to 1000 c.c. NaHO.

(*d*) $\underbrace{2NaHO}_{2)80} + \underbrace{H_2SO_4}_{2)98} = Na_2SO_4 + 2H_2O$

40 = 49 = grms. of H_2SO_4, equivalent to 1000 c.c. NaHO.

(*e*) $\underbrace{2NaHO}_{2)80} + \underbrace{H_2C_4H_4O_6}_{2)150} = Na_2C_4H_4O_6 + 2H_2O$

40 75 = grms. of $H_2C_4H_4O_6$, equivalent to 1000 c.c. NaHO.

(*f*) $\underbrace{3NaHO}_{3)120} + \underbrace{H_3C_6H_5O_7 \cdot H_2O}_{3)210} = Na_3C_6H_5O_7 + 4H_2O$

40 70 = grms. of $H_3C_6H_5O_7 \cdot H_2O$, equivalent to 1000 c.c. NaHO.

(*C*) List of Strengths of B.P. Acids taken by Soda.

Acetic (glacial)	99% $HC_2H_3O_2$.
,, (ordinary)	33% ,,
,, (dilute)	4·27% ,,
Citric	pure.
Hydrochloric	32% HCl.
,, (dilute)	10·58% ,,
Lactic	75% $HC_3H_5O_3$.
,, (dilute)	13% ,,
Nitric	70% HNO_3.
,, (dilute)	17·44% ,,
Nitro-hydrochloric	32 grms. should neutralise 88·3 c.c. NaHO
Sulphuric	98% H_2SO_4
,, (dilute)	13·65% ,,
,, (aromatic)	12·5% ,,
Tartaric	99% $H_2C_4H_4O_6$.
Hydrobromic	10% HBr.

IV. STANDARD SOLUTION OF ARGENTIC NITRATE.

$\frac{N}{10}$ = 17 *grammes per* 1000 *c.c.*

(*A*) Preparation.

Dissolve 17 grammes pure $AgNO_3$ in distilled water, and make up to 1 litre (1000 c.c.)

Check.—As argentic nitrate is not always pure, the solution when thus made should be standardised by weighing out ·1 (one decigramme) of *pure* powdered sodium chloride, dissolving it in water in a small beaker, adding sufficient solution of potassium chromate to color it yellow, and then running in the silver solution, with constant stirring, until the last drop just causes the color to change from yellow to pink. This should take 17·1 c.c. of silver solution if it be of correct strength, because:—

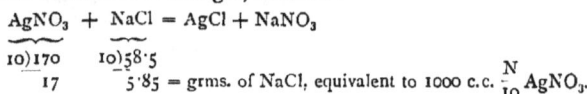

$$\underbrace{AgNO_3}_{10)170} + \underbrace{NaCl}_{10)58{\cdot}5} = AgCl + NaNO_3$$

$$17 \qquad 5{\cdot}85 = \text{grms. of NaCl, equivalent to 1000 c.c. } \frac{N}{10}\,AgNO_3.$$

(*B*) Estimation of Soluble Haloid Salts.

The silver solution is used for the estimation of haloid salts by weighing out any quantity ranging between ·1 and ·2 (one or two decigrammes), dissolving and titrating, K_2CrO_4 being used as the indicator, exactly as above described. Thus we should operate upon:—

(*a*) *Potassium bromide.*

$$\underbrace{AgNO_3}_{10)170} + \underbrace{KBr}_{10)119} = AgBr + KNO_3$$

$$17 = 11{\cdot}9 = \text{grms. of KBr equivalent to 1000 c.c } \frac{N}{10}\,AgNO_3.$$

(*b*) *Ammonium bromide.*

$$\underbrace{AgNO_3}_{10)170} + \underbrace{NH_4Br}_{10)98} = AgBr + NH_4NO_3$$

$$17 \qquad 9{\cdot}8 = \text{grms. of } NH_4Br \text{ equivalent to 1000 c.c. } \frac{N}{10}\,AgNO_3.$$

(*c*) *Sodium bromide, potassium iodide, and sodium iodide,* all by similar equations.

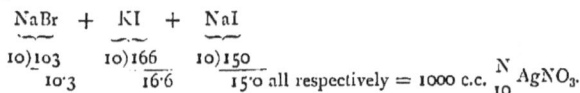

$$\underbrace{NaBr}_{10)103} + \underbrace{KI}_{10)166} + \underbrace{NaI}_{10)150}$$

$$10{\cdot}3 \qquad 16{\cdot}6 \qquad 15{\cdot}0 \text{ all respectively} = \text{1000 c.c. } \frac{N}{10}\,AgNO_3.$$

Note.—Bromides, if adulterated with iodides, will take *less* silver than they ought, but, if the impurity be chloride, they will take *more*. Therefore they must neither take *less* nor *more* than the correct amount.

(*C*) Application of Volhard's Method to the Analysis of Chlorides.

This process is, in certain cases, a much better method for the estimation of chlorides than the direct method with the chromate indicator, because it may be used in the presence of nitric acid, thus enabling a chloride to be estimated in presence of a phosphate or other acid which precipitates silver in a neutral solution. It depends upon entirely precipitating the chloride in the presence of nitric acid by a known volume of standard solution of silver, and then estimating the excess of silver, left uncombined with the chloride, by standard solution of ammonium thiocyanate (sulphocyanate) using a drop of solution of ferric alum as an indicator. The thiocyanate solution is checked against the silver so as to correspond with it c.c. to c.c. As soon as the thiocyanate has precipitated all the silver it will strike a blood-red with the ferric salt due to the formation of ferric thiocyanate. The difference between the volume of standard silver solution originally added and that of the thiocyanate used will give the c.c. of silver equivalent to the chloride present.

(*D*) Estimation of Hydrocyanic Acid.

Silver solution is also used for taking the strength of hydrocyanic acid. This is done by weighing out about 5 grammes, adding a drop of solution of litmus, and then a few drops of NaHO so as to cause strong alkalinity. Now carefully drop in the silver solution from the burette, with constant agitation, until a *faint permanent cloud* of AgCN is produced. If, during the operation, the litmus becomes red, more NaHO must be added to keep it blue. At the moment that the *permanent* cloud appears the following reactions are complete :—

$$\underbrace{2HCN}_{54} + 2NaHO = \underbrace{2NaCN}_{98} + 2H_2O$$

$$\underbrace{2NaCN}_{98} + \underbrace{AgNO_3}_{170} = AgCN \cdot NaCN + NaNO_3$$

therefore—

$$\underbrace{2HCN}_{10)54} = \underbrace{AgNO_3}_{10)170}$$

$$5{\cdot}4 = 17{\cdot}0$$

so that 1000 c.c. $\frac{N}{10}$ $AgNO_3$ are equivalent to 5·4 grms. of HCN.

On adding more silver AgCN would be precipitated thus :—

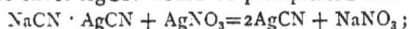

$$NaCN \cdot AgCN + AgNO_3 = 2AgCN + NaNO_3;$$

and the first appearance of this precipitate therefore shows that the action to be calculated upon is complete.

Potassium cyanide is done in the same way, using about ·2 gramme, and $AgNO_3$=2KCN—*i.e.*, 17 grammes (1000 c.c.)=6·5 grammes KCN.

(*E*) Standards of Strength by Silver Solution.

Hydrocyanic acid	2%.
Potassium cyanide	95%.
Potassium iodide	98 to 99%.
Sodium iodide	99%.
Potassium bromide	98 to 100%.
Sodium bromide	99%.
Ammonium bromide	pure.

V. STANDARD SOLUTION OF IODINE.

$\frac{N}{10}$ = 12·7 *grammes per litre* (1000 *c.c.*)

(*A*) Preparation.

Weigh out 12·7 grammes of pure iodine and place it in a litre flask with 18 grammes of potassium iodide and about 200 c.c. of water, agitate till dissolved, and make up to one litre with water.

Check.—To standardise the strength of the solution (if desired), test it against ·2 gramme (2 decigrammes) of pure As_2O_3, as hereafter described.

(*B*) Estimation of Arsenious Acid.

This solution is used :—

For *arsenious acid* and the solutions thereof weigh out one to two decigrammes of the As_2O_3, and dissolve it in boiling water by the aid of five times its weight of $NaHCO_3$. Let it cool, add some mucilage of starch, and run in the iodine solution until a faint permanent blue color is obtained. Then apply the equation :—

$$\underbrace{2I_2}_{4)508} + \underbrace{As_2O_3}_{4)198} + 5H_2O = 2H_3AsO_4 + 4HO$$

$$\begin{array}{ll} 4)508 & 4)198 \\ 10)127 & 10)49{\cdot}5 \\ \quad 12{\cdot}7 & \quad 4{\cdot}95 \text{ grms. of } As_2O_3 \text{, equivalent to 1000 c.c. } \frac{N}{10} \text{ iodine.} \end{array}$$

For *liquor arsenicalis* use 10 grammes and ·3 gramme $NaHCO_3$.
For *liquor arsenici hydrochloricus* use 10 grammes and ·6 gramme $NaHCO_3$.

(*C*) Estimation of Sulphurous Acid.

For *sulphurous acid* weigh out about one gramme from a stoppered bottle, and largely dilute it with water. Add starch mucilage, and run in the iodine solution until the faintest possible *permanent* blue appears. Then apply the equation :—

$$\underbrace{I_2} + H_2O + \underbrace{SO_2} \cdot H_2O = H_2SO_4 + 2HI$$

$$\begin{array}{ll} 2)254 & 2)64 \\ 10)127 & 10)32 \\ \quad 12{\cdot}7 & \quad 3{\cdot}2 \text{ grms. } SO_2 \text{, equivalent to 1000 c.c. } \frac{N}{10} \text{ iodine.} \end{array}$$

(*D*) Estimation of Thiosulphates.

For *sodium thiosulphate* (hyposulphite) use about ·5 gramme, dissolve in water, add starch, and titrate as above.

12·7 I=24·8 "hypo."

VI. STANDARD SOLUTION OF SODIUM THIOSULPHATE ("HYPO.")

Strength $\frac{N}{10}$ = 24·8 "*hypo*" ($Na_2S_2O_3 \cdot 5H_2O$) *per litre* (1000 *c.c.*)

(*A*) Preparation and Check.

Dissolve about 28 grammes of commercial "hypo" in a litre of distilled water. Now place 20 c.c. of standard solution of iodine slightly diluted with water in a beaker, and run in the rough "hypo" solution until the color changes to that of pale sherry. Then add mucilage of starch, and continue to run in the "hypo" till the last drop *just discharges the blue* color. Note the number of c.c. of "hypo" used, put 50 times that number into a test mixer, and make up with distilled water to 1000 c.c.

This solution may be used for :—

(*B*) Estimation of Free Iodine.

By weighing out ·2 gramme (2 decigrammes), dissolving in a little water by the aid of potassium iodide, and then running in "hypo" till the color is reduced to that of a pale sherry; lastly, adding starch mucilage, and going on till the blue is just bleached. Then by the equation:—

$$\underbrace{2(Na_2S_2O_3 5H_2O)}_{\substack{2)496\\10)248\\24\cdot8}} + \underbrace{I_2}_{\substack{2)254\\10)127\\12\cdot7}} = 2NaI + Na_2S_4O_6 + 10H_2O$$

24·8 = 12·7 grms. I, equivalent to 1000 c.c. $\frac{N}{10}$ "hypo."

(C) Estimation of Free Chlorine or Bromine.

For *chlorine water or bromine water.*

Weigh about 10 grammes from a stoppered bottle, pouring it directly into a flask containing 2 grammes of potassium iodide dissolved in 50 c.c. of water, and then titrate with "hypo" as already described.

The Cl first liberates an equivalent quantity of iodine from the KI, and the "hypo" then acts upon the I_2 so set free, thus:—

$$\underbrace{Cl_2}_{\substack{20)71\\3\cdot55}} + 2KI = 2KCl + \underbrace{I_2}_{\substack{20)254\\12\cdot7}}$$

$$\underbrace{2Na_2S_2O_3 5H_2O}_{\substack{20)496\\24\cdot8}} + \underbrace{I_2}_{\substack{20)254\\12\cdot7}} = 2NaI + Na_2S_4O_6 + 10H_2O$$

24·8 = 12·7, therefore 3·55 grms. Cl = 1000 c.c. $\frac{N}{10}$ "hypo."

On the same principle,

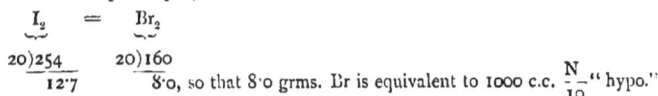

$$\underbrace{I_2}_{\substack{20)254\\12\cdot7}} = \underbrace{Br_2}_{\substack{20)160\\8\cdot0}}$$

8·0, so that 8·0 grms. Br is equivalent to 1000 c.c. $\frac{N}{10}$ "hypo."

(D) Estimation of Available Chlorine.

For *chlorinated lime* and its solution.

Use ·5 gramme of the solid or 5 grammes of the *liquor*. Put it into a flask with 1·5 gramme of KI dissolved in 100 c.c. of water, and then drop in HCl in excess. Lastly, titrate with "hypo," as already described. Then by the equations:—

(*a*) $CaO\,Cl_2 + 2HCl = CaCl_2 + H_2O + Cl_2$,
(*b*) $Cl_2 + 2KI = 2KCl + I_2$,
(*c*) $2(Na_2S_2O_3 \cdot 5H_2O) + I_2 = 2NaI + Na_2S_4O_6 + 10H_2O$,

we come to the result already shown for chlorine-water—namely, that

$$\underbrace{2(Na_2S_2O_3 . 5H_2O)}_{\substack{20)496\\24\cdot8}} = \underbrace{I_2}_{\substack{20)254\\12\cdot7}} = \underbrace{Cl_2}_{\substack{20)71\\3\cdot55}}$$

therefore, 1000 c.c. $\frac{N}{10}$ = "hypo" represent 3·55 grammes "*available chlorine*" in all chlorinated compounds.

Liquor sodæ chlorinatæ.

Use about 5 grammes, and proceed as for chlorinated lime. The action and calculations are the same, only differing in the first equation, which is

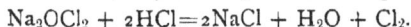

$$Na_2OCl_2 + 2HCl = 2NaCl + H_2O + Cl_2.$$

(*E*) List of B. P. Standards by Iodine and "Hypo."

Iodine sol.		"Hypo" sol.	
Sulphurous acid .	. 5% SO_2	Chlorine water . . .	·6% Cl.
Arsenious acid .	. pure.	Chlorinated lime . . .	33% Cl.
Arsenious liquors .	. 1% As_2O_3	Liquor calcis chlor. . .	2 to 3% Cl.
Sodium hyposulphite .	pure	Liquor sodæ chlor. . .	2·5% Cl.

VII. STANDARD SOLUTION OF POTASSIUM BICHROMATE.

Strength $\frac{N}{20}$ = 14·75 *grammes* $K_2Cr_2O_7$ *in* 1000 *c.c.*

(*A*) Preparation.

Dissolve 14·75 grammes pure $K_2Cr_2O_7$ in one litre of water.

Check.—If desired to standardise, we do so by performing an estimation upon ·5 gramme (5 decigrammes) of pure iron (pianoforte) wire, dissolved in dilute H_2SO_4 as described below.

Principle of the process.—$K_2Cr_2O_7$, when heated with an acid (say H_2SO_4), gives

$$K_2Cr_2O_7 + 4H_2SO_4 = K_2SO_4 + Cr_2(SO_4)_3 + 4H_2O + O_3;$$

therefore, each *molecule* of bichromate gives *three atoms* of nascent oxygen, which possesses the power of raising *six atoms* of iron from the *ferrous* to the *ferric state*, because

$$6FeO + O_3 = 3Fe_2O_3.$$

(*B*) Estimation of Ferrous Salts.

Bichromate solution is to be used for the following estimations:—

(*a*) *Metallic iron.*

·5 gramme of the metallic wire or filings is dissolved by the aid of heat in dilute sulphuric acid, using a flask fitted with a cork through which passes a small tube to allow the exit of steam and hydrogen, but to prevent as far as possible ingress of air. While the iron is dissolving, 2 ounces of water are placed in a basin over the gas, and a burette charged with bichromate solution is arranged over it. A porcelain slab is also got ready at the right-hand side of the basin, and is covered over with spots, from a glass rod, of freshly made solution of potassium ferricyanide. When dissolved, the iron solution is rinsed from the flask into the basin, and immediately titrated with the bichromate, until a drop taken from the basin on the stirring rod just ceases to give a blue color when brought into contact with one of the spots of potassium ferricyanide on the slab. The number of c.c. of bichromate solution used is read off, and the following equations applied:—

$$\underbrace{6Fe}_{20)336 = 16·8} + 6H_2SO_4 = \underbrace{6FeSO_4}_{20)912 = 45·6} + 3H_2$$

$$\underbrace{6FeSO_4}_{20)912 = 45·6} + \underbrace{K_2Cr_2O_7}_{20)295 = 14·75} + 7H_2SO_4 = 3Fe_2(SO_4)_3 + K_2SO_4 + Cr_2(SO_4)_3 + 7H_2O;$$

therefore 14·75 $K_2Cr_2O_7$ = 16·8 Fe grammes, equivalent to 1000 c.c. $\frac{N}{20}$ bichromate.

(*b*) For ferrous sulphate, ferrous phosphate, ferrous arseniate, and *ferri carbonas saccharata*.

Use about 1 gramme in each case, dissolve in water by the aid of sulphuric or hydrochloric acid, then add a good excess of dilute H_2SO_4 or HCl, and titrate with bichromate solution as already described. Apply the equations as under:—

(1) *Crystallised ferrous sulphate.*

$$\underbrace{K_2Cr_2O_7}_{20)295} + \underbrace{6FeSO_4 \cdot 7H_2O}_{20)1668} + 7H_2SO_4 = 3Fe_2(SO_4)_3 + K_2SO_4 + Cr_2(SO_4)_3 + 14H_2O$$

14·75 83·4 = grms., equivalent to 1000 c.c. $\frac{N}{20}$ bichromate.

(2) *Real ferrous sulphate.*

$$\underbrace{K_2Cr_2O_7}_{20)295} + \underbrace{6FeSO_4}_{20)912} + 7H_2SO_4 = 3Fe_2(SO_4)_3 + K_2SO_4 + Cr_2(SO_4)_3 + 7H_2O$$

14·75 45·6 = grms., equivalent to 1000 c.c. $\frac{N}{20}$ bichromate.

(3) *Ferrous phosphate.*

$$\underbrace{K_2Cr_2O_7}_{20)295} + \underbrace{2Fe_3(PO_4)_2 . 8H_2O}_{20)1004} + 7H_2O + 7H_2SO_4 = Fe_2(SO_4)_3 + 2Fe_2(PO_4)_2 + [K_2SO_4 + Cr_2(SO_4)_3 + 15H_2O$$

14·75 50·2 = grms., equivalent to 1000 c.c. $\frac{N}{20}$ bichromate.

(4) *Ferrous arseniate.*

$$\underbrace{K_2Cr_2O_7}_{20)295} + \underbrace{2Fe_3(AsO_4)_2}_{20)892} + 7H_2O + 7H_2SO_4 = Fe_2(SO_4)_3 + 2Fe_2(AsO_4)_2 + [K_2SO_4 + Cr_2(SO_4)_3 + 7H_2O$$

14·75 44·6 = grms., equivalent to 1000 c.c. $\frac{N}{20}$ bichromate.

(5) *Ferri carb. sacch.* (dissolve in HCl, because H_2SO_4 would char the sugar):—

$$\underbrace{K_2Cr_2O_7}_{20)295} + \underbrace{6FeCO_3}_{20)696} + 26HCl = 3Fe_2Cl_6 + 2KCl + Cr_2Cl_6 + 6CO_2 + 13H_2O$$

14·75 34·8 = grms., equivalent to 1000 c.c. $\frac{N}{20}$ bichromate.

(*C*) Estimation of Ferric Salts.

The process may be applied to *hæmatite* or any form of *ferric* iron by dissolving in and acidulating with hydrochloric acid, and then dropping in zinc so as to evolve H_2 and reduce the iron to the ferrous state. When all Zn has dissolved, and the solution has ceased to give a red with KCNS, it is titrated with bichromate in the usual way.

(*D*) B.P. Standards by Bichromate Solution.

Iron	pure.
Crystallised ferrous sulphate	54·13% real $FeSO_4$ (=99% crystals).
Granulated ,, ,,	54·67% real $FeSO_4$ (=100% crystals).
Dried ,, ,,	87·0% real $FeSO_4$.
Ferrous phosphate	47% $Fe_3(PO_4)_2$, $8H_2O$.
Ferrous arseniate	10% $Fe_3(AsO_4)_2$
Ferrous carbonate sacch.	*about* 33% $FeCO_3$

Instead of bichromate we may use a standard solution of potassium permanganate to estimate iron, simply running it in (in presence of excess of

H_2SO_4) until a permanent pink color remains. Each molecule of $K_2Mn_2O_8$ gives O_5, and therefore equals ten atoms of Fe in oxidation.

VIII. FEHLING'S STANDARD SOLUTION OF COPPER.

(*A*) Manufacture and Check by the Ordinary Method.

This solution is used for the estimation of sugars, and is made as follows :— Dissolve 34·65 grammes of pure crystallised sulphate of copper in 250 c.c. water. Dissolve 173 grammes rochelle salt and 60 grammes KHO in 500 c.c. water. Mix the two solutions, and make up, when quite cold, to 1000 c.c. Each 10 c.c. of this liquid will represent—

Glucose	·05 gramme.
Maltose	·082 "
Inverted cane sugar	·0475 "
Inverted starch	·045 "

To check Fehling's solution, weigh out ·475 gramme of pure sugar candy and dissolve it in 100 c.c. of water in a small flask; add 3 drops of strong HCl, and boil briskly for ten minutes to invert the cane sugar into glucose. Let it cool, neutralise with KHO, and then make up exactly to 100 c.c. with distilled water. Place this liquid in a burette arranged over a basin placed over the gas, and containing 10 c.c. of Fehling's solution and 50 c.c. of water. When the contents of the basin are boiling, run in the sugar solution until all blue color is destroyed. Then note the number of c.c. of sugar solution used, and whatever that number may be, it will contain the equivalent in sugar of 10 c.c. of "Fehling." If the "Fehling" be correct, 10 c.c. of the standard sugar will be used to entirely precipitate it.

It is usually necessary to do the estimation twice, first roughly and then accurately, using the second time drops of K_4FeCy_6 acidulated with acetic acid, on a slab, as an indicator for the disappearance of the last trace of Cu from solution.

(*B*) Manufacture and Check by Pavy's Method.

Cuprous oxide dissolves in ammonia, forming a colorless liquid. Taking advantage of this point, Pavy treats an ammoniacal cupric solution at a boiling temperature with sufficient saccharine solution to exactly discharge the blue color. The advantage of this method over that above described lies simply in the fact that there is no bulky red precipitate to interfere with the ready observation of the end reaction. To prepare the test solution, dissolve 20·4 grms. of rochelle salt and the same weight of caustic potash in distilled water; dissolve separately 4·158 grms. of pure cupric sulphate in more water with heat; add the copper solution to that first prepared, and when cold, add 300 c.c. of strong ammonia, and distilled water to 1 litre. The process is conducted as follows: 10 c.c. of the ammoniated cupric solution (=0·005 grm. of glucose) are diluted with 20 c.c. of distilled water, and placed in a small flask. This is attached by means of a cork to the nozzle of a burette, fitted with a glass stopcock, and previously filled with the saccharine solution previously diluted to a fixed bulk. The cork of the flask should be traversed by a small bent tube, to permit steam to escape. Now heat the flask until the blue liquid boils. Turn the stopcock in order to allow the saccharine solution to flow into the hot solution—which should be kept at the boiling point—at the rate of about 100 drops per minute (not more nor much less), until the azure tint is exactly discharged. Then stop the flow, and note the number of cubic centimetres used. That amount of saccharine solution will contain 5 milligrammes of glucose. To render the determination as accurate as possible, the solution for analysis should be diluted to such an extent that not less than 4 nor more than 7 c.c. are required to decolorise the solution.

To find the percentage of sugar, multiply 0·005 by the original total bulk (in c.c.) of the solution started with, and divide the product by the number of cubic centimetres of solution used from the burette. To observe easily the exact end-reaction, a piece of paper or other white body should be placed behind the flask. Mr. Stokes uses the half of an ordinary opal gas globe fixed in the proper position. If the operator objects to the escape of the waste ammoniacal fumes, they may be conducted by a suitable arrangement into water or dilute acid. For a special apparatus for this purpose see the *Analyst*, vol. xii.

(*C*) Estimation of Sugar.

The sugar weighed must not exceed ·5 gramme, and must be dissolved in 100 c.c. of water. If the sugar be either glucose or maltose or lactose it is titrated directly; but if cane sugar, it is first inverted as above described. By always placing 10 c.c. of "Fehling" in the basin, then whatever number of c.c. of sugar solution we use, that number will contain the equivalent of 10 c.c., and we have only to calculate:—

As No. of c.c. used : Total volume of sugar solution :: Equivalent of 10 c.c. "Fehling" of the sugar in question : Real sugar present in the quantity weighed out for analysis.

(*D*) Estimation of Starch.

Starch is weighed and boiled in a flask with water containing dilute hydrochloric acid, under an upright condenser, for some hours. It is then cooled, neutralised with potassium hydrate, diluted to a fixed volume (not stronger than 1 in 200), and then the solution so made is titrated into 10 c.c. of "Fehling." A much improved process will be found in Chapter X.

IX. ESTIMATION OF PHOSPHORIC ACID

is performed by means of a standard solution of uranic nitrate in the presence of sodium acetate. The necessary solutions are, —

1. Standard solution of uranic nitrate, made by dissolving 70 grammes in 900 c.c. of water, and then, after ascertaining its strength by performing an analysis on 50 c.c. of the standard phosphate solution, diluting with water so that 50 c.c. will correspond exactly to 50 c.c. of that solution. If absolutely pure uranic nitrate were obtainable, theory requires the solution of 71 grammes in one litre of water to yield a solution which will balance the standard phosphate (each 1 c.c. = 1 gramme of P_2O_5).
2. **Standard phosphate solution,** made by dissolving 50·42 grammes of perfectly pure disodium hydrogen phosphate in one litre of water, when each 1 c.c. will equal ·1 gramme of P_2O_5.
3. A solution of 100 grammes of sodium acetate and 100 grammes of acetic acid in water, and the whole diluted to one litre.
4. Finely powdered potassium ferrocyanide.

To perform the process, the solution of the phosphate in about 50 c.c. of water is placed in a basin on the water bath, mixed with 5 c.c. of solution No. 3 (sodic acetate), and No. 1 (uranic nitrate), is run in from a burette, until a drop taken from the basin on to a white plate just gives a brown color, when a little powdered ferrocyanide is cautiously dropped into its centre. The number of c.c. of uranic solution used having been noted, the usual calculations are to be applied.

After repeated trials upon 50 c.c. of the standard phosphate solution, so as to thoroughly adjust the strength of the uranic solution, and at the same time accustom the eye to observe the exact moment of the appearance of the brown coloration, the process may be practically applied to **Manures.**

The best method of preparing the solution of the manure is to heat 10 grammes to dull redness for 15 minutes, and when cold to reduce it to a fine powder in a mortar, and add gradually 10 grammes of sulphuric acid diluted to 200 c.c. with water. Rinse the whole into a stoppered bottle, and make up with water to one litre. Shake up occasionally for an hour, and having then let all settle for three hours, draw off 100 c.c. (=1 gramme manure) for analysis. To this add a little citric acid (10 drops of a cold saturated solution), and slightly supersaturate with ammonium hydrate. Again acidify with acetic acid, add 10 c.c. sodium acetate solution, and then use the uranic solution as usual. If all these quantities be rigorously adhered to, each c.c. of uranic solution used can, without further calculation, be taken as indicating 1 per cent. of tricalcium phosphate in the manure.

This process is highly recommended by Mr. Sutton of Norwich, and elaborate details will be found in his work on Volumetric Analysis.

X. STANDARD SOLUTION OF BARIUM CHLORIDE.

Semi-normal = 104 *grammes per* 1000 *c.c. of* $BaCl_2$.

This is used for taking the amount of a soluble sulphate, by adding it to a known weight of the sulphate; dissolve in water acidulated with hydrochloric acid, until precipitation ceases. The process, however, is tedious, and the end of the reaction is not sharp, and it is therefore rarely employed. The following is a specimen of the reaction using magnesium sulphate:

$$MgSO_4 \cdot 7H_2O + BaCl_2 = BaSO_4 + MgCl_2 + 7H_2O.$$

Each c.c. of the standard solution equals ·04 SO_3 or ·048 SO_4.

The solution is made by dissolving 104 grammes of pure barium chloride dried at 220° F. in 1 litre of water.

XI. STANDARD MAYER'S SOLUTION.

Made by dissolving 13·546 grammes of pure mercuric chloride and 49·8 grammes of potassium iodide in water, and then making up to 1000 c.c.

This solution is used for the estimation of alkaloids, which should be free from any mucilaginous matter and preferably dissolved in a little dilute sulphuric acid. The reagent is added till precipitation ceases. In the author's hands the process has not worked very well, except for the amount of emetine in *ipecacuanha*, which may be rapidly ascertained as follows:—

15 grammes of ipecacuanha are treated with 1·5 c.c. of dilute sulphuric acid, and sufficient rectified spirit of wine added to make the whole bulk up to 150 grammes. The whole is allowed to stand for 24 hours, and 100 c.c. are decanted off for analysis. The liquid is evaporated until all the alcohol is driven off, and then brought under the burette containing the test solution, which is run in until it ceases to give a precipitate. The final point of the reaction is ascertained by filtering off a drop or two into a watch-glass placed on black paper, and adding a drop of the reagent, when, if no cloudiness appear, the precipitation of the alkaloid is complete. The number of c.c. of the test used multiplied by ·0189 gives the amount of alkaloid in 10 grammes

of the sample, which again multiplied by 10 gives percentage. A full list of the equivalents of the various alkaloids to Mayer's solution is found in the "coefficient" table at the end of this chapter.

XII. ANALYSIS BY THE NITROMETER.

(*A*) General Remarks.

This useful instrument is illustrated in fig. 29. It consists of a measuring tube (A) graduated in cubic centimetres, having a funnel-shaped cup (C) connected to it by means of the stopcock (D). This cock is a "three-way" one, and according to the direction in which it is turned, it can make connection and discharge the contents of the cup either into the tube A or out in the waste opening at E; or it can make, or quite shut off, all connection between (A) and the outer air through E. Connected to A by a piece of flexible indiarubber tube is the ungraduated control tube B. The object of the apparatus is the rapid and accurate measurement, at definite temperature and pressure, of gases evolved during any reaction; and it takes its name from the fact that it was first used to measure the nitric oxide given off by the decomposition of nitric acid. Suppose that we fill the instrument with a fluid (say mercury) right up to the top: having closed the tap D, we lower the tube B and then admit a little carbonic anhydride through E; by opening and again closing the top, we have a volume of gas in the measuring tube which we desire to measure under definite condition. If (1) we allow the instrument to stand until its contents must have assumed the temperature of the room, then a *centigrade* thermometer suspended to the same stand will give the temperature of the gas; (2) If we now raise or lower the control tube so that the level of the liquid both in it and in the measuring tube is the same, it is evident that the pressure inside A is the same as in the room, and reference to a barometer standing near will give that pressure. It now only remains to read off the volume of the gas in the measuring tube, and having corrected it to normal temperature and pressure by Charles's and Boyle's laws respectively, to calculate it from its volume in c.c. to its weight in grammes by multiplying the number of c.c. of volume at N.T.P. by the weight of 1 c.c. of the gas in grammes. This latter is easily obtained by multiplying the *crith* (·0896 gramme, weight of 1 litre of H) by the *atomic* weight of an *elementary* or *half* the *molecular* weight of a *compound* gas, and then dividing by 1000. Suppose, for example, that in the analysis of a nitrate or nitrite we have obtained 20 c.c. of nitric oxide at 15° C. and 750 m.m. barometer, and we require to know the weight of NO so got, that we may afterwards calculate therefrom the weight of the nitrate present, we should say :—

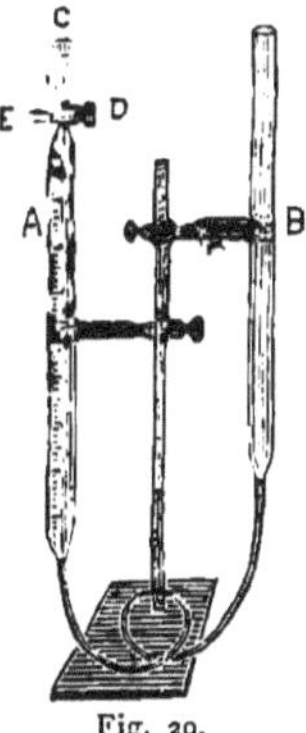

Fig. 29.

(*a*) $\frac{(273 + 0) \times 750 \times 20}{(273 + 15) \times 760} = 18{\cdot}788$ c.c., corrected volume at N.T.P.

(*b*) $\frac{{\cdot}0896 \times 15}{1000} = {\cdot}001344$ grm., weight of 1 c.c. NO.

(*c*) $18{\cdot}788 \times {\cdot}001344 = {\cdot}0253$ grm., weight of NO found.

The various possible applications of this instrument are so numerous that an exhaustive detail would be impossible in the present work; but the following should be practised as typical instances of its use :—

(*B*) Estimation of the Strength of Spirit of Nitrous Ether.

The active principle of this drug is ethyl nitrite. Nitrites when mixed with excess of potassium iodide and acidulated with sulphuric acid cause a liberation of iodine and evolve all their nitrogen in the form of nitric oxide, thus :—

$$C_2H_5 \cdot NO_2 + KI + H_2SO_4 = C_2H_5 \cdot HO + KHSO_4 + I + NO.$$

The process (originally due to Mr. Allen) has been recognised officially for the assay of the spirit, and is thus conducted. The nitrometer is filled with saturated solution of sodium chloride, with which owing to its density a strong spirit will not readily mix, and so we save the expense of mercury; 5 c.c. of the sample to be tested is placed in the cup of the nitrometer, and the control tube having been lowered, the spirit is allowed to enter through the tap, taking care that no air gets in at the same time. 5 c.c. of a strong solution of potassium iodide is next allowed to enter, and this is in turn followed by 5 c.c. of dilute sulphuric acid. Effervescence immediately occurs, and if the tube be vigorously agitated at intervals, the reaction completes itself in 10 minutes. The level of the liquid in the control tube of the instrument is adjusted to equal that in the measuring tube, and the volume of nitric oxide is read off and calculated. For official purposes it is however sufficient to see that the resulting gas is seven times (or at all events not less than five times) the volume of the spirit started with. The *least trace* of air allowed to enter with the liquids vitiates the results of the process, because the nitric oxide would be thereby converted into a higher oxide of nitrogen and so become soluble in the fluid with which the instrument is charged.

(*C*) Estimation of Nitric Acid in Nitrates.

This depends on the fact that when a nitrate is shaken up with excess of sulphuric acid and mercury the following reaction takes place :—

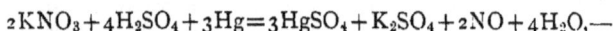

$$2KNO_3 + 4H_2SO_4 + 3Hg = 3HgSO_4 + K_2SO_4 + 2NO + 4H_2O,—$$

thus showing that each molecule of the nitrate radical NO_3 gives off a molecule of NO. If any chlorides or other haloid salts be present they are first removed by adding a slight excess of argentic sulphate to the solution and filtering. No quantity of a nitrate exceeding 2 decigrammes should be used, otherwise more gas may be evolved than the instrument will conveniently hold. The nitrometer is charged with mercury, and the nitrate solution, which should not exceed 5 c.c., is put into the cup and passed therefrom into the measuring tube followed by excess of strong sulphuric acid. The instrument is well agitated for some time, and when action has ceased and the contents have cooled down to the temperature of the room, the level is adjusted and the volume of NO read off and calculated. All the precautions already mentioned must be observed. If any nitrites be present, they affect the accuracy of the estimation, being also decomposed to nitric oxide.

(*D*) Estimation of a Soluble Carbonate.

This has been proposed for use in taking the strength of the medicinal solution of ammonium carbonate in the spirit known as *spiritus ammoniæ aromaticus*. A given volume of the spirit is placed in the cup and introduced into the nitrometer charged with mercury. This is followed by an excess of dilute hydrochloric acid, and the carbon dioxide evolved by the action of the acid upon the carbonate is measured. From this the percentage of ammonium carbonate may be calculated, or an empirical comparison of volume on the principle of that already described for spirit of nitrous ether may be applied. According to Mr. Gravill, the originator of the test, good aromatic spirit of ammonia should give off seven times its volume of carbon dioxide after allowing a correction for the slight solubility of the gas in the liquid with which it is inclosed.

(*E*) Estimation of the Strength of Solutions of Hydrogen Peroxide.

This depends upon the fact that, when hydrogen peroxide acts upon potassium permanganate, acidulated with sulphuric acid, oxygen is evolved. One half of this oxygen is due to the peroxide and the other to the permanganate. The nitrometer should be charged with concentrated solution of sodium sulphate (which in this case is better than brine), and one cubic centimetre of the solution is introduced from the cup followed by excess of a strong solution of potassium permanganate acidulated with sulphuric acid. The contents of the measuring tube, after the reaction is complete, must remain colored violet, thus showing that sufficient permanganate has been employed. A solution commercially described as of 10 per cent. strength by volume should when thus treated give off twenty times its volume of oxygen.

(*F*) Estimation of Urea in Urine.

This process depends on the fact that when urea is decomposed by an alkaline hypobromite or hypochlorite, it gives off its nitrogen in the free state. A small flask is fitted with a tight cork through which passes a funnel tube closed by a clamp and reaching to the bottom of the flask and also a bent delivery tube just passing through the cork. 5 c.c. of the urine is placed in this flask, and the nitrometer having been filled with water the flask is attached to the tap of the nitrometer at the end E (*see* fig. 29, page 122). An excess of a solution of bromine in sodium hydrate solution is then placed in the funnel and allowed to run into the urine, and the clamp immediately closed. At the same moment the tap of the nitrometer is so placed as to establish connection between E and the measuring tube. A little warm water in a basin is applied to the flask to hasten the reaction, and when no more gas is evolved, the tap is closed, the temperature and pressure adjusted, and the volume read off as usual. Each c.c. of gas at N.T.P. represents ·0029 gramme of urea present in the 5 c.c. of urine acted upon.

XIII. COLORIMETRIC ANALYSIS.

General Remarks.

This is a variety of volumetric analysis in which the amount of a substance present in solution is found by adding, to a given volume, a fixed quantity of a reagent and observing the color produced. This color is then matched by adding, to an equal volume of distilled water, the same mixed quantity of reagent, and running in a volumetric solution of the pure substance until the same tint is produced. Evidently when this point is reached the amount of substance present in the solution under analysis equals that in the volumetric solution used for the comparative experiment. The applications commonly occurring of this method are:

(*A*) Estimation of Ammonia by "Nesslerising."

For this process the following solutions and apparatus are required:—

(*a*) *Nessler's solution.* Dissolve 35 parts of potassium iodide in 100 parts of water. Dissolve 17 parts of mercuric chloride in 300 parts of water. The liquids may be heated to aid solution, but if so must be cooled. Add the latter solution to the former until a permanent precipitate is produced. Then dilute with a 20 per cent. solution of sodium hydrate to 1000 parts; add mercuric chloride solution until a permanent precipitate again forms; allow to stand till settled, and decant off the clear solution. The bulk should be kept in an accurately stoppered bottle, and a quantity transferred from time to time to a small bottle for use. The solution improves by keeping.

(*b*) *Standard ammonia solution.* Dissolve 3·146 grammes pure ammonium chloride in 1000 c.c. of distilled water free from ammonia. For use dilute 10 c.c. of this solution to 100 c.c. with ammonia-free distilled water. Each c.c. of the diluted solution will then contain ·1 milligramme of NH_3 (*i.e.* ·0001 gramme).

(*c*) *Two narrow cylinders of colorless glass*, of perfectly equal height and diameter, holding about 150 c.c., and graduated at 100 c.c. These should either have a milk glass foot or should stand upon perfectly white paper.

(*d*) *A pipette* to deliver $1\frac{1}{2}$ c.c.

(*e*) *A quantity of ammonia-free distilled water.* This is obtained by placing a litre of ordinary distilled water in a retort attaching a condenser (see Chap. I., page 4), and distilling until what passes over ceases to give any color with "Nessler's solution." The remaining water in the retort is then cooled and bottled for use.

The liquid in which ammonia is to be estimated (usually a distillate obtained in water analysis) is first made up to a fixed bulk, and the bulk noted. It must be so diluted that it only gives a *color* and not a *precipitate* with "Nessler." 100 c.c. of this solution are placed in a cylinder, and $1\frac{1}{2}$ c.c. of "Nessler" having been added by the pipette, and the whole stirred with a *perfectly clean rod*, the color produced is observed. A little experience soon teaches the operator to judge the probable amount of ammonia solution required to produce a similar tint. Let us suppose, for example, that the color is judged to be equal to 2 c.c. of ammonia, then we proceed to confirm our idea: 2 c.c. of the standard ammonia solution are run from a burette into the other cylinder, ammonia-free water is added to 100 c.c., then the $1\frac{1}{2}$ c.c. of "Nessler," and the whole stirred with the clean rod. If now, after a few minutes, the colors match, we are correct; but if not, then we must try again and again with more or less standard ammonia until we get an exact match between the colors in the two cylinders. This having been attained, the calculation is very simple, and will be best explained by an example. Suppose that we start with a distillate containing ammonia and made up to 200 c.c., and that we employ 5 c.c. of standard ammonia solution in the comparison experiment, to match the color produced by "Nessler" in 100 c.c.

of such distillate. Then 5 c.c × ·1 = ·5, and ·5 × 2 = 1·0; therefore the whole 200 c.c. of distillate contained 1 milligramme of NH_3. Beginners should train their eyes by observing the colors produced by adding various quantities of standard ammonia to 100 c.c. of ammonia-free water and then introducing the "Nessler." $\frac{1}{10}$ of a c.c. of standard ammonia will produce a very faint yellow, while larger amounts will increase the color to orange, and finally to deep orange-red. We should always wait 3 minutes before observing, as the full color does not appear under that time.

(*B*) Estimation of Salicylic Acid.

The substance containing the salicylic acid is slightly acidulated with hydrochloric acid, and then shaken up with ether in a separator. The ether is drawn off and evaporated, and the residue is dissolved in water and diluted to a fixed bulk until it gives a suitable violet color, with a few drops of ferric chloride. 100 c.c. of the solution are then treated with 10 drops of solution of ferric chloride; and the color having been observed, it is imitated in the other cylinder by a very dilute standard solution of pure salicylic acid made to 100 c.c. with distilled water and a similar number of drops of ferric chloride, exactly like the nesslerising process above described.

XIV. TABLE OF COEFFICIENTS REQUIRED IN THE VOLUMETRIC ANALYSIS DESCRIBED IN THIS CHAPTER.

Standard Acid Solution.	COEFF.
Oxalic acid	·063
Sulphuric acid	·049
NH_3	·017
NH_4HO	·035
$NH_4HCO_3 . NH_4NH_2CO_2$	·0523
$Na_2B_2O_7 . 10H_2O$	·191
$Pb(C_2H_3O_2)_2 3H_2O$	·1895
$Pb_2O(C_2H_3O_2)_2$	·137
Ca2HO	·037
CaO	·028
$CaCO_3$	·05
$Ba(HO)_2$	·0855
$Ba(HO)_2 8H_2O$	·1575
$BaCO_3$	·0985
MgO	·02
$MgCO_3$	·042
KHO	·056
K_2CO_3	·069
K_2CO_3 + 16 °/₀ Aq	·08214
$K_2C_4H_4O_6$ (converted into K_2CO_3)	·113
$(K_2C_4H_4O_6)_2H_2O$	·1175
$KHC_4H_4O_6$	·188
$K_3C_6H_5O_7$	·102
$KC_2H_3O_2$	·098
$K_2Mn_2O_8$ (by oxalic acid)	·0314
$KNaC_4H_4O_6$	·141
NaHO	·04
Na_2CO_3	·053
$Na_2CO_3 . 10H_2O$	·143
$NaHCO_3$	·084

Standard Soda Solution.	
Sodium hydrate	·04
$HC_2H_3O_2$	·06
$H_3C_6H_5O_7H_2O$	·07
HCl	·0365
HBr	·0808
HI	·128
HNO_3	·063
H_2SO_4	·049
$H_2C_4H_4O_6$	·075
$HC_3H_5O_3$ (lactic)	·09
$H_2C_2O_4 2H_2O$	·063

Standard Nitrate of Silver Solution.	
Argentic nitrate	·017
CN	·0052
HCN	·0054
KCN	·01302
NH_4Cl	·00535
KCl	·00745
NaCl	·00585
KBr	·0119
NaBr	·0103
Na_2HAsO_4	·0064
Cl	·00355

Standard Iodine Solution.	COEFF.
Iodine	·0127
SO_2	·0032
H_2SO_3	·0041
As_2O_3	·00495
$Na_2S_2O_3 5H_2O$	·0248
$Na_2SO_3 7H_2O$	·0126
$K_2SO_3 2H_2O$	·0097

Standard Bichromate of Potassium.	
Potassium bichromate	·01475
Fe(Ferrous)	·0168
$FeSO_4$	·0456
$FeSO_4 2H_2O$	·051
$FeSO_4 7H_2O$	·0849
$Fe_3(AsO_4)_2$	·0446
$Fe_3(PO_4)_2$	·0358
$FeCO_3$	·0348
Fe_3O_4	·0696
FeO	·0216

Standard Hyposulphite Solution.	
Hyposulphite of sodium	·0248
I	·0127
Cl	·00355
Br	·0080

Standard Barium Chloride Solution.	
Barium chloride ($BaCl_2$)	·104
$BaCl_2 2H_2O$	·122
H_2SO_4	·049
SO_4	·048
SO_3	·040

Standard Mayer's Solution.	
Aconitine	·0267
Atropine (1 · 200)	·0097
Nepaline	·0388
Hyoscyamine (1–200)	·00598
Emetine	·0189
Coniïne	·0125
Nicotine	·00405
Strychnine	·0167
Brucine	·0197
Colchicine (1–600)	·0317
Morphine	·02
Narcotine	·0213
Veratrine	·0296
Sabadilline	·0374
Sabatrine	·03327
Physostigmine	·01375
Berberine	·0425

Nitrometer Analysis.

Each cc. of NO at N.T.P. equals gramme of	·00282 HNO_3 ·00262 N_2O_5 ·00452 KNO_3 ·00336 $C_2H_5 . NO_2$ ·0029 Urea
Each cc. of CO_2 at N.T.P. equals gramme of	·0043 $(NH_4)_2CO_3$

CHAPTER VIII.

GRAVIMETRIC QUANTITATIVE ANALYSIS OF METALS AND ACIDS.

DIVISION I.—PRELIMINARY REMARKS.

GRAVIMETRIC quantitative analysis is that method by which the substance to be estimated is converted into some chemically definite compound, weighed as such, and the amount of the original substance obtained from this weight by calculation. The same definite compound will answer both for the estimation of its metal and of its acid. For example, if we precipitate a known weight of argentic nitrate with hydrochloric acid we obtain insoluble argentic chloride which may be filtered out and weighed and the amount (x) of Ag in the quantity started with calculated therefrom, because,—

AgCl : Ag : : weight of AgCl found : x.
143·5 : 108 : : „ „ „ : x.

If, on the other hand, we start with a known weight of hydrochloric acid, precipitate it with argentic nitrate, and collect and weigh the argentic chloride formed, we can find the amount (x) of real HCl actually present in quantity started with, because,—

AgCl : HCl : : weight of AgCl found : x.
143·5 : 36·5 : : „ „ „ : x.

Before giving the individual processes for the quantitative estimation of the various metals, we must first say something about the preparation of filters, and the washing, drying, and weighing of precipitates, which will serve as general directions and so save continual repetition of details.

(*A*) The Preparation of Filters.

Ready-cut filters may be procured from the dealers in chemical apparatus. The kind known as Swedish is the best for all cases where the precipitate is finely divided or pulverulent. For gelatinous precipitates, such as ferric hydrate and calcium phosphate, the white English filters are preferable; but they should never be used, say, for barium sulphate or calcium oxalate, as those bodies would very likely pass through the pores of the filter, and so cause a loss in the analysis. The only drawback to Swedish papers is their filtering rather slowly. Whatever paper be used, the size for quantitative operations is, for the larger sort, six inches in diameter, for the smaller, about two inches. The small sort is used where we have to deal with traces of precipitate only, or when a small quantity of fluid has to be filtered. The paper should yield nothing to dilute acids, and if the ash exceed one milligramme per large filter it will in most cases be remedied by placing, say, 100 cut filters for some hours in a basin filled with a mixture of one volume of HCl and eight volumes of water. They must then be repeatedly washed with distilled water till quite free from acidity, otherwise they would crumble to pieces when being folded. The washing is a very tedious operation indeed, and having been completed, the basin is put on to a water bath till the filters are perfectly dry.

(*B*) Estimation of the Ash of Filters.

This is most conveniently done by folding ten filters in a small compass, twisting a long platinum wire round the packet so as to form a cage, holding the free end in the hand and the paper over a previously weighed platinum crucible while touching it with the flame of a Bunsen burner. The paper burns and the ash drops into the crucible, while any particles of carbon which have escaped combustion are quite consumed by exposing the crucible for some time to a red heat till the ash gets perfectly white. The crucible after cooling is reweighed, and its increase is the ash of ten filters. Divided by 10, we get the ash of one filter; and in every case where both filter and precipitate are burned, the ash of the filter thus found must always be deducted from their total weight, and the difference is then the actual weight of the precipitate.

(*C*) The Collection and Washing of Precipitates.

When the precipitate has been fully formed and the supernatant fluid has become quite clear, the latter is poured on the filter (which is either previously tared or not according to circumstances), care being taken not to disturb the precipitate. This is done by holding a glass rod in a perpendicular position over the filter, placing the lip of the beaker against it and causing the liquid to flow steadily down the rod into the filter. When the latter is three-fourths full, the beaker is turned into an erect position and the rod drained against the inside of the lip and then laid across the top of the beaker until it is time to refill the filter. After thus pouring off as much as practicable, the precipitate remaining in the beaker is treated with water and well stirred. When the whole has once more settled, the clear fluid is again passed through the filter. This operation having been repeated three or four times, the precipitate is allowed to pass on to the filter, any particles which stick to the sides of the beaker being removed with a feather, or a rod tipped with a small piece of black india-rubber tubing; and the whole having been thus collected, the washing is continued by means of a washing-bottle (fig. 30), till the precipitate is quite freed from its soluble impurities. For instance, in estimating sulphuric acid, the barium sulphate is washed till the filtrate no longer gives a turbidity with argentic nitrate.

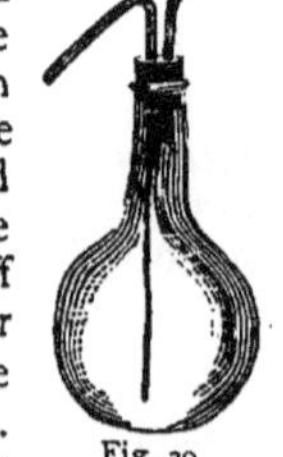

Fig. 30.

Many bodies, as ferric and aluminic hydrate, most phosphates, barium sulphate, and some of the carbonates, are best washed with boiling water. Others, on the contrary, must be washed with cold water,—such as plumbic sulphate, for which we use cold water acidified with some H_2SO_4; magnesium ammonium phosphate, for which cold dilute ammonium hydrate is used, etc.

(*D*) Drying of Precipitates.

After the precipitate has been thoroughly washed and perfectly drained, the funnel containing it is loosely covered over with filter paper and then put into the water oven or air bath till dry. Most precipitates are dried at a temperature of 212° F., but some of them require a heat of 220° F. before becoming constant in weight. Prolonged and repeated drying is only necessary when the precipitate is weighed on the filter, as described below. Fig 31 shows a water oven for drying at 212° F., while fig. 32 shows an air bath for drying at higher temperatures. This bath is fitted with an apparatus for automatically controlling the gas supply, and consequently the temperature.

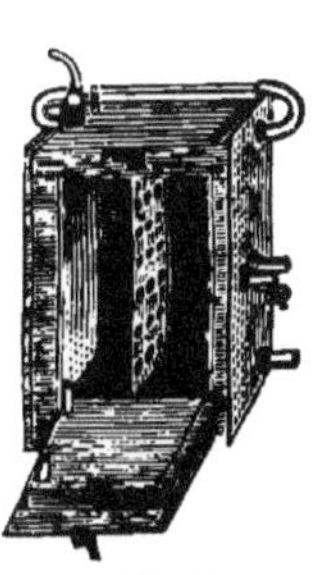

Fig. 31.

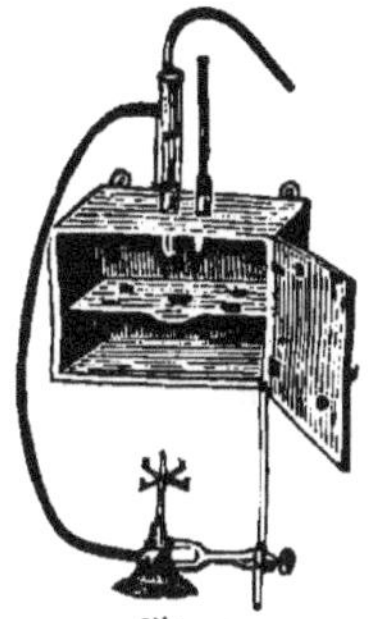

Fig. 32.

(*E*) Igniting and Weighing of Precipitates.

Most precipitates must first be ignited before they can be weighed. This is to drive off water, which they may still retain after drying at 212° F., or carbonic anhydride and water. For instance, zinc is best weighed as oxide, and therefore the precipitate, consisting of oxycarbonate, is first ignited. Iron is precipitated as hydrate, but the composition of that body not being constant, it is ignited and so made into pure oxide before weighing. As soon, therefore, as the precipitate appears dry, it is carefully detached from the filter and put into a previously ignited and weighed crucible, the filter is burned on the lid (which has been weighed together with the crucible), the ash is thrown into the crucible, and the latter covered with the lid. The crucible is now supported by a pipe-clay triangle, and gently ignited at first, to prevent spurting from the sudden evolution of steam or other gases. The lid is now taken off, and the crucible inclined a little, so as to give a free access of air. The ignition is continued for some minutes, and the crucible, having been again covered with the lid, is allowed to cool and weighed. For accurate estimations it is best to let it cool under a desiccator. Such an arrangement is shown in fig. 33, in which A is a vessel containing strong sulphuric acid to keep the air under the glass shade always free from moisture.

Fig. 33.

Fig. 34.

The heat of an ordinary Bunsen burner is generally sufficient for all purposes; but the conversion of calcium carbonate into oxide requires the aid of a gas blowpipe; while argentic chloride must be heated with a rose Bunsen or spirit lamp until it just begins to fuse. The filters are, as already shown, burned separately, to prevent any reduction of the precipitate by the carbon of the filter.

Some precipitates are not ignited but weighed on a previously tared filter. Before weighing the filter—for which purpose a weighing tube (fig. 34) is used, or the filter is placed between two closely-fitting watch-glasses provided with a clamp to hold them together—it must first be dried for fifteen minutes at 212° F. After drying the precipitate, the filter is again placed in the tube or between the glasses and reweighed: the increase shows, of course, the weight of the substance. It is well to replace the filter in the bath for, say, half an hour, and to weigh again. Should the weight be considerably less, it must be once more put into the bath and reweighed. Another method of weighing precipitates on a filter is to prepare two filters of equal size, A and B. Cut off the bottom point of B, so that A will go inside it with its point projecting through the opening. Now put B on the weight scale of the balance, and cut off minute slices from the top of A until the two are exactly counterbalanced. Place A inside B, and then, having put both in the funnel, collect the precipitate, wash, and dry as usual, and cool under the desiccator. Lastly, detach B, and use it for a tare, putting it into the weight pan, and then the weights required to balance A and its contents will be the weight of the precipitate, because both filters having been exposed to the same conditions the tare is accurate.

(*F*) Estimation of Moisture.

A watch-glass is exactly tared on the balance, and then 2 grammes of the substance (in powder if possible) are carefully weighed upon the glass, and the total weight noted. The glass, with contents, is then placed in the drying oven and heated therein for an hour, at the expiration of which it is removed to the desiccator, and when cold, is weighed and the weight noted. It is then replaced in the oven for half an hour, and the cooling and weighing repeated. If the two weights do not agree within, say, 2 milligrammes, the process is repeated until two concordant weighings are obtained. The weight after drying deducted from the total weight of glass + substance started with gives the moisture, which figure multiplied by 50 gives percentage.

(*G*) Estimation of the Ash of Organic Bodies.

This determination is necessary in every analysis of a vegetable or animal substance. A platinum dish is heated to redness, cooled under the desiccator, weighed, and the weight noted. A suitable quantity, say 5 to 10 grammes of the substance, is weighed into the dish, which is then arranged on a triangle support over a Bunsen burner and heated to dull redness. If after fumes cease the substance is seen to have assumed a coke-like form, it is removed from the dish into a small dry mortar, and having been carefully powdered, the powder is replaced in the dish and maintained at a dull red heat until it has become perfectly white, or at least until all carbon has been burned off. If the burning proves very tedious and the last traces of carbon are very difficult to burn, the addition of a light sprinkling of ammonium nitrate will cause the process to complete itself more rapidly. The dish is now cooled under the desiccator and weighed, and the weight of the empty dish having been deducted, the difference gives the weight of the ash, which is then calculated to percentage. The heat should not be allowed to rise to bright redness, because potassium and sodium chlorides, which are very common constituents of the ash, would be thereby volatilised to some extent and so lost.

(*H*) Analytical Factors for Calculating the Results of Analyses.

To save the working out of a rule of three sum on the result of each analysis it is customary to employ factors. These are obtained by dividing the weight of the required body by the equivalent weight of the form in which it is precipitated. Thus, supposing we are estimating the amount of argentic nitrate present in a solution containing ·6 gramme of the salt, and have precipitated and weighed the same in the form of argentic chloride, we have :—

$$\frac{\text{Molecular weight of } AgNO_3 \; 170}{\text{Equivalent weight of } AgCl \; 143{\cdot}5} = 1{\cdot}1847, \text{ analytical factor.}$$

It now only remains to multiply this factor by the weight of the precipitate to obtain the answer. Let us further suppose that the weight of the precipitate was ·5 gramme, then $1{\cdot}1847 \times {\cdot}5 = {\cdot}59235$ real $AgNO_3$ present in the ·6 gramme taken; then $\frac{{\cdot}59235 \times 100}{{\cdot}6} = 98{\cdot}73$ per cent. real $AgNO_3$ present in the sample.

DIVISION II. GRAVIMETRIC ESTIMATION OF METALS.

1. ESTIMATION OF SILVER.

(*A*) As Argentic Chloride.

(Practise upon ·5 gramme pure $AgNO_3$ dissolved in 100 c.c. H_2O.)

Silver is most conveniently weighed as chloride, because this body is perfectly insoluble in water and dilute acids, and separates readily. The silver solution to be estimated is acidified with nitric acid, and hydrochloric acid is dropped in until no more precipitate forms. It is best to have the solution warm, and to stir till the supernatant liquid has got perfectly clear. The clear fluid is now poured off through a filter, and the chloride is washed by decantation with boiling water, always pouring the washings through the filter, till every trace of acid is removed, and subsequently brought upon the filter. The filter and contents is then dried in the water oven, and the chloride transferred into a weighed porcelain crucible, and heated till it just commences to fuse. The filter is burned on the crucible lid, and the ash treated with a drop of aqua regia, the resulting chloride dried, the lid placed upon the crucible, and the whole weighed. The tare of the crucible and lid having been deducted, the balance is AgCl, from the quantity weighed out for analysis.

(*B*) As Metal.

If silver is required to be estimated in organic salts, and other bases are absent, the analysis is performed by igniting a weighed quantity of the salt, and weighing the ash, which will consist of pure metallic silver.

In alloys the pure metal is obtained by *cupellation*, as follows :—The weighed alloy is wrapped in lead foil, placed on a little cup or *cupel* made of bone ash, and heated to bright redness in a muffle furnace. The lead oxidises and sinks into the cupel, carrying the impurities with it, and leaving a button of pure silver, which is cooled and weighed.

2. ESTIMATION OF LEAD.

(*A*) As Plumbic Oxide.

(Practise upon ·5 gramme pure plumbic acetate.)

The solution containing the substance to be analysed is precipitated with ammonium carbonate in the presence of a little ammonium hydrate. The precipitated plumbic carbonate is then collected on a small filter, washed, and dried. The dry precipitate is removed as completely as possible from the filter-paper, and introduced into a weighed porcelain crucible. The filter having been burned on the lid, and its ash added to the contents of the crucible, the whole is ignited, cooled, and weighed. By ignition, the oxide is formed; and after deducting the weight of the crucible, the balance is PbO, from the quantity weighed out for analysis. Organic salts of lead require simply to be ignited in a tared porcelain crucible, with free access of air, adding towards the end a sprinkling of ammonium nitrate, and weighing the resulting PbO.

(*b*) As Plumbic Chromate.

(Practise upon ·5 gramme of plumbic nitrate.)

The solution is mixed with excess of sodium acetate and precipitated with potassium chromate. The precipitate is collected, washed with water, acidulated with acetic acid, dried and ignited in a platinum crucible with the usual precautions. The filter is burned on the lid, treated with a drop or two of nitric acid, dried and again ignited. The crucible and contents are weighed, and the tare of the crucible having been deducted, the balance is $PbCrO_4$, from the quantity weighed out for analysis.

Lead may also be precipitated as $PbSO_4$, dried, ignited and weighed as such.

3. ESTIMATION OF MERCURY.

(*a*) As Metal.

(Practise upon 1 gramme of "white precipitate," which should yield 77·5 % Hg.)

Take a combustion tube of hard glass closed at one end, and put in: (1) a little magnesite—$MgCO_3$; (2) the white precipitate mixed with excess of quick-lime; (3) a few inches of quick-lime; (4) a loose plug of asbestos. Draw out the open end of the tube before the blowpipe and bend it down at a right angle. Give the tube a tap or two on the table to insure a free passage along the upper part for gases, and place it in a combustion furnace, with its open end dipping under the surface of some water in a small flask. Apply heat to the front part of the tube and go gradually backwards until the whole is heated to redness and the CO_2, given off by the $MgCO_3$, has swept all the mercury vapor out of the tube. The mercury collects as a globule under the water in the basin, and is transferred to a tared watch-glass, perfectly dried by pressure with blotting-paper, and weighed.

(*b*) As Mercuric Sulphide.

(Practise upon ·5 gramme of mercuric chloride.)

Through the solution of the mercuric salt a current of H_2S is passed till the liquid is saturated. It is necessary the mercury should be all in the mercuric state, and not contain other metals precipitable by H_2S. The precipitate is collected on a weighed filter, washed first with water, then with absolute alcohol, and finally, to remove any free sulphur, with a mixture of equal parts of ether and carbon disulphide. After drying at 212° F. and weighing, the balance is HgS from the quantity weighed out for analysis.

4. ESTIMATION OF CADMIUM.

(Practise upon ·5 gramme of $CdCO_3$ dissolved in diluted HCl.)

As Sulphide.

The solution is precipitated with ammonium hydrate and ammonium sulphide. The cadmium sulphide is collected in a weighed filter, washed, dried at 212° F., and weighed. This process is only applicable in the absence of metals also precipitable by ammonium sulphide. In the presence of metals of the fourth group the solution must be slightly acidified with hydrochloric acid and precipitated by a current of sulphuretted hydrogen.

5. ESTIMATION OF COPPER.

(*a*) As Cupric Oxide.

(Practise upon ·5 gramme of pure $CuSO_4·5H_2O$.)

The solution (freed from other metals if necessary) is boiled with a slight excess of sodium hydrate. The precipitate is filtered out, washed, and dried. It is then carefully removed from the paper to a weighed crucible, and the filter having been burned on the lid and the ash added to the contents of the crucible, the whole is well ignited, cooled in a desiccator, and weighed rapidly, because cupric oxide is very hygroscopic. To make sure that the oxide contains no cuprous oxide, it is moistened with a little fuming nitric acid, dried with the lid on, ignited for ten minutes, and then reweighed. This operation requires care, being liable to involve loss by spurting.

(*b*) As Metallic Copper.

The solution, which must be free from metals precipitable by electrolysis, is introduced into a weighed and very clean platinum basin. It must contain a slight excess of sulphuric acid, but on no account nitric acid. The dish is then attached to the wire from the zinc plate of a galvanic cell, thus becoming the *cathode*. The other wire is connected to a piece of platinum wire to form an *anode*, and this latter is then immersed in the liquid. After a short time the fluid will become quite colorless, and the basin will be coated with metallic copper. The fluid is now poured off, and the copper repeatedly washed with boiling water till all acidity is removed. The basin is finally rinsed with absolute alcohol, quickly dried, weighed, and the tare of the basin having been deducted the difference is metallic copper in the quantity weighed out for analysis.

The use of a battery may be dispensed with and a fragment of pure zinc used to precipitate the copper on the basin with sufficient acid to dissolve all the zinc before pouring off.

6. ESTIMATION OF BISMUTH.

(*a*) As Bismuth Oxide.

The solution for analysis (freed, if necessary, from other metals) is diluted with water, and precipitated with a slight excess of ammonium carbonate. The precipitated bismuthous oxycarbonate is collected, washed and dried. It is then separated from the filter paper, and the latter having been burned on the lid of a weighed crucible, the whole is introduced into the crucible, and ignited, cooled, and weighed.

(*b*) As Bismuth Sulphide.

(Practise upon ·75 gramme of *bismuthi et ammoniæ citras* B. P.)

This process (although employed in the B.P.) cannot be much recommended, as the sulphide is apt to increase in weight on drying, owing to the absorption of oxygen. A current of sulphuretted hydrogen is passed through the acid bismuth solution ; the resulting sulphide is collected on a tared filter, dried at 212° F., and weighed.

7. ESTIMATION OF GOLD.

As Metallic Gold, by Cupellation.

The alloy containing the gold is treated exactly as described under silver. The resulting metallic button is rolled out into a flat foil, and is then digested with nitric acid, which dissolves any silver, and the resulting gold is re-fused into a button and weighed.

8. ESTIMATION OF PLATINUM.

As Metallic Platinum.

The solution, which must contain the platinum as chloride, is concentrated and precipitated with excess of ammonium chloride. The precipitate is well washed with spirit of wine, dried, ignited, and weighed, as metallic platinum.

9. ESTIMATION OF TIN.

(*a*) As Stannic Oxide.

Alloys containing tin, but free from antimony or arsenic, are treated with nitric acid, which converts the tin into oxide and other metals into nitrates. The acid fluid is evaporated nearly to dryness, the residue taken up with water and a little nitric acid ; the oxide is washed by decantation, collected on a filter, completely washed and dried. It is then as completely as possible detached from the filter, the latter is burned on a lid, the ash added to the contents of the crucible, and the whole ignited. After cooling the oxide is moistened with a little nitric acid, dried (with the lid on), and again ignited, when, after cooling, it is ready for weighing.

Where we have to deal with tin in solution, the following method is applied :—

The solution, which must be free from metals of the first three groups, is precipitated with sulphuretted hydrogen, the resulting sulphide is washed with solution of ammonium acetate, which will prevent the stannic sulphide from passing through the filter. After drying, the sulphide is transferred to a weighed crucible, the filter burned on the lid, its ash added to the contents of the crucible, and the whole ignited, *at first very gently*, until fumes of sulphurous anhydride cease, and then at a very high temperature, with the addition of a fragment of ammonium nitrate.

This process depends on the conversion of the sulphide into SnO_2 by ignition ; but it must be conducted with care, as a too rapid application of heat would cause the change to take place suddenly, and the whole would ignite.

(*b*) As Metallic Tin.

This process, which is only applicable to tin stone, consists in fusing a known quantity of the pulverised ore with potassium cyanide in a porcelain crucible, when a small button or granules of metallic tin will be obtained on treating the mass with water. The tin is washed, dried, and weighed.

10. ESTIMATION OF ANTIMONY.

As Antimonious Sulphide, with or without Subsequent Conversion into Antimonious Antimonic Oxide.

(Practise upon ·5 gramme of "tartar emetic.")

The acid solution is mixed with tartaric acid, to prevent the precipitation of an oxysalt, diluted with water, and precipitated with sulphuretted hydrogen, the sulphide collected on a weighed filter, dried at 220° F., and weighed. The conversion of the sulphide into oxide is done by igniting it in a porcelain crucible, heating gently at first; then dropping on nitric acid until red fumes cease, and finally igniting very strongly. The remaining Sb_2O_4 is then weighed.

11. ESTIMATION OF ARSENIC.

(*a*) As Arsenious Sulphide.

(Practise upon ·5 gramme As_2O_3.)

The solution must contain the arsenic as arsenious acid. After adding some HCl, a current of sulphuretted hydrogen is passed through the liquid till the latter acquires a strong smell. The excess of gas is now removed by warming the fluid and passing a current of carbonic anhydride through it. The sulphide is collected on a weighed filter, washed, dried at 212° F., and weighed.

(*b*) As Magnesium Ammonium Arseniate.

If the substance be arsenious acid it is dissolved in some hot solution of sodium carbonate, excess of hydrochloric acid is added, and the fluid *gently* heated with potassium chlorate till it smells distinctly of chlorine, even after half an hour. Arsenic and sulphur compounds, on the other hand, are dissolved in hot potassium hydrate and treated with excess of chlorine gas to convert them into arsenic acid. The solution of arsenic acid thus obtained by either of the foregoing methods is mixed with large excess of ammonium hydrate, and, after being allowed to cool, precipitated with *magnesia mixture* (a fluid prepared by adding ammonium hydrate to a solution of magnesium sulphate and re-dissolving the precipitate with ammonium chloride). After standing for at least twelve hours, the precipitate is collected on a weighed filter, washed with a mixture of one volume of ammonium hydrate and three volumes of water till free from chlorine, dried for three hours at 220° F. and weighed. If only dried at 212° F. the drying will occupy a considerably longer time. The results are usually a trifle too low.

12. ESTIMATION OF COBALT.

As Potassium Cobaltous Nitrite.

The solution is concentrated to a small bulk, the excess of acid is neutralised with potash, and excess of potassium nitrite and a little acetic acid (to keep the solution slightly acid to test-paper) are then added. After the lapse of twenty-four hours all the cobalt will have crystallised out as potassium cobaltous nitrite. This salt is quite insoluble in the mother liquor, but slightly so in pure water. For the washing, a 10% solution of potassium acetate is used, wherein the salt is also insoluble, and the acetate is afterwards removed by washing with alcohol. A weighed filter is used and the precipitate dried at 212°.

13. ESTIMATION OF NICKEL.

As Metal.

The solution is precipitated with excess of sodium hydrate and boiled. The precipitate is washed with boiling water, dried, ignited, and weighed. The ignited residue, or a known portion of it, is now introduced into a weighed glass tube and reduced at red heat by a current of hydrogen. The reduced metallic nickel is afterwards weighed.

14. ESTIMATION OF MANGANESE.

As Manganoso-manganic Oxide.

(Practise upon ·75 gramme of pure $MnSO_4 \cdot 7H_2O$.)

The solution for analysis, if strongly acid, is neutralised with ammonium hydrate and precipitated by ammonium sulphide. The precipitated manganous sulphide is washed with water containing ammonium sulphide, and dissolved in acetic acid. Chlorine gas is passed through the liquid until all the manganese precipitates as manganic peroxide, which is then collected, washed, and calcined in a weighed crucible to bright redness This forms Mn_3O_4; the crucible with the contents is then cooled and weighed.

15. ESTIMATION OF ZINC.

As Zinc Oxide.

(Practise upon ·75 gramme of pure $ZnSO_4 \cdot 7H_2O$.)

The solution of the zinc salt (freed from other metals if necessary) is precipitated boiling with sodium carbonate, and the solution boiled well. The precipitate is allowed to settle, washed by decantation with boiling water, filtered out, and dried. It is then introduced into a weighed crucible, ignited for some time at a bright red heat, cooled, and weighed. The ignition changes the precipitated zinc carbonate to oxide, and it is weighed as such. Or the solution is precipitated with ammonium sulphide, the zinc sulphide collected on a filter, washed with dilute ammonium sulphide, dried, ignited, and finally weighed as oxide. This latter method is useful when only small quantities of zinc are present.

16. ESTIMATION OF IRON.

As Ferric Oxide.

(Practise upon ·75 gramme of pure $FeSO_4 \cdot 7H_2O$.)

The solution (freed if necessary from other metals, phosphates, and organic matter) is boiled with a few drops of nitric acid, to insure that the whole of the iron is in the ferric state. Excess of ammonium hydrate is added, the whole boiled, and rapidly filtered. The precipitated ferric hydrate is washed with boiling water, dried, and ignited in a weighed crucible for some time.

In the presence of organic matter, such as citric or tartaric acids, the iron must first be separated by precipitation with ammonium sulphide, the precipitate washed with dilute ammonium sulphide, redissolved in hydrochloric acid, boiled, oxidised by potassium chlorate, and then precipitated with ammonium hydrate, as directed. In the presence of manganese the solution should be nearly neutralised by ammonium hydrate and then boiled with excess of ammonium acetate, and the resulting ferric oxy-acetate collected, washed, dried, and ignited to Fe_2O_3.

17. ESTIMATION OF ALUMINIUM.

As Aluminic Oxide.

(Practise upon 1 gramme of pure alum.)

The solution containing the alum or other salt of the metal (which must be free from iron and earthy phosphates) is precipitated with a slight excess of ammonium hydrate, and boiled until it only smells very faintly of ammonia. The precipitated aluminic hydrate thus obtained is filtered out, washed with boiling water, and dried. The dry filter and its contents are transferred to a weighed platinum crucible, and ignited to bright redness for some time, allowed to cool, and weighed.

18. ESTIMATION OF CHROMIUM.

As Chromic Oxide.

Salts of chromium are at once precipitated with ammonium hydrate, and the precipitate washed, dried, ignited, and weighed as Cr_2O_3. Soluble chromates are first reduced by means of hydrochloric and sulphurous acids (or, instead of the latter, spirit of wine may be used), and the chromium precipitated as hydrate by ammonium hydrate ignited and weighed as Cr_2O_3, all as described above for aluminium.

19. ESTIMATION OF BARIUM.

As Barium Sulphate.

(Practise upon ·5 gramme of $BaCl_2 . 2H_2O$.)

To a solution in boiling water add excess of sulphuric acid, boil rapidly for a few minutes, and set aside to settle.

The clear liquor is poured off as closely as possible, and the precipitate collected on a filter of *Swedish paper*, and washed with boiling water. The filter and precipitate are next dried and ignited in a weighed platinum crucible (the precipitate being removed as perfectly as possible from the paper, and the latter first burned separately on the crucible lid, and the ash added to the contents of the crucible, to avoid the reduction of $BaSO_4$ to BaS by the carbon of the paper). The crucible and its contents having been weighed, and the weight of the crucible deducted, the difference equals the $BaSO_4$.

20. ESTIMATION OF CALCIUM.

As Calcium Carbonate or Calcium Oxide.

(Practise upon ·5 gramme of powdered calc-spar dissolved in dilute HCl.)

The solution of the lime salt is mixed with ammonium chloride, and is then made alkaline by ammonium hydrate. Should any precipitate (for instance, calcium phosphate) form, it is redissolved by means of acetic acid, and any insoluble residue is removed by filtration. Ammonium oxalate is now added in excess. The precipitated calcium oxalate is boiled for a few minutes, filtered, the precipitate washed until free from oxalates, and the filter and contents dried at 212° F. The precipitate is now carefully transferred to a tared platinum crucible and gently ignited. It is then moistened with a solution of pure ammonium carbonate, evaporated to dryness, heated until no more fumes are evolved, and then weighed as carbonate.

21. ESTIMATION OF MAGNESIUM.

As Magnesium Pyrophosphate.

(Practise upon ·5 gramme of pure $MgSO_4 \cdot 7H_2O$.)

The solution, which must be strong, is mixed with some ammonium chloride, and then with one-third of its bulk of ammonium hydrate, entirely cooled, excess of di-sodium hydrogen phosphate added, and the whole set aside for some hours. Care must be taken not to touch the sides of the beaker with the stirring rod, as otherwise particles of the triple phosphate will adhere to them so tenaciously that they can only be removed with great difficulty. The precipitate is collected on the filter and washed with a mixture of one volume of ammonium hydrate and three volumes of water, till the washings are free from chlorine, and dried. The precipitate is now detached from the filter and put into a weighed platinum crucible, the filter is burned in the lid, the ash added to the contents of the crucible, and the whole ignited by means of a Bunsen burner; and, when a large quantity of phosphate has been obtained, finally ignited before the blowpipe.

It sometimes happens that the phosphate, even after prolonged ignition, is very black. In that case it is, after cooling, thoroughly moistened with nitric acid, carefully dried, and re-ignited, when it will be found to be perfectly white.

22. ESTIMATION OF POTASSIUM.

As Potassium Platino-Chloride.

(To be practised upon ·2 gramme of pure KCl.)

The solution is placed in a small porcelain basin, and mixed with a *good* excess of solution of platinic chloride. The whole is evaporated to dryness on a water bath kept at a temperature of about 200° F. When quite dry it is again digested with a few drops of platinic chloride solution, to make sure that all the sodium will be in the state of sodium platino-chloride. This sodium compound and the excess of platinic chloride are now removed by spirit of wine of 60 O.P., the precipitate is collected on a weighed filter, washed with alcohol till the washings appear quite colorless, dried at 212° F., and weighed.

23. ESTIMATION OF SODIUM.

As Sodium Sulphate.

(Practise upon ·5 gramme of pure NaCl.)

This method is applicable where we have to deal with a sodium salt containing a volatile acid. The solution is mixed with excess of sulphuric acid and evaporated in a weighed platinum basin. When fumes of sulphuric acid become visible, the basin is covered over with a lid or foil which has been weighed together with the crucible, and gradually heated till fumes cease. While red-hot the foil is lifted up a little, and a small lump of ammonium carbonate put in the crucible, which operation is repeated after a few minutes, and the residual Na_2SO_4 is cooled and weighed. The object of introducing the ammonium carbonate is to remove the last traces of free sulphuric acid.

24. ESTIMATION OF POTASSIUM AND SODIUM IN PRESENCE OF METALS OF FOURTH GROUP.

(Practise upon the residue left on evaporating 1 litre of ordinary drinking-water, and redissolved in a little dilute HCl.)

The solution, which must be free from other metals, but may contain calcium, magnesium, and sodium compounds (as their presence does not interfere with the process), is first of all precipitated with excess of barium chloride, which throws down sulphuric, phosphoric, etc., acids. Barium hydrate or some milk of lime is now added in slight excess, when any magnesia will also be thrown down. To the filtered liquid excess of ammonium carbonate is added, the precipitate is separated by the filter and the fluid evaporated in a platinum crucible to dryness, best on the water bath. When quite dry it is covered with a platinum lid and gently heated as long as white ammoniacal fumes are visible. The residue, which will now consist of potassium chloride, together with perhaps sodium chloride, is however not quite fit for weighing, and must be purified. This is done by redissolving in water and adding a little ammonium carbonate, when a slight precipitate will form. After filtering, the fluid is evaporated, this time in a weighed platinum basin covered with a weighed lid, on the water bath, and when dry the residue is gently heated to faint redness for a minute, cooled and weighed. When no sodium is present it will now be pure potassium chloride; but should it also contain sodium chloride, then it must be redissolved, the potassium estimated by $PtCl_4$, and the sodium obtained by difference. This process is one of the most commonly occurring operations, because it is required in every full analysis of water, and also of the ash of all vegetable and animal substances, where both potassium and sodium have always to be estimated in presence of Ca, Mg, phosphates, etc. It is therefore a very important one to thoroughly master.

25. ESTIMATION OF AMMONIUM.

As Ammonium Platino-Chloride.

If the solution contains other basylous radicals, a known quantity of it is distilled with some slaked lime in a suitable apparatus, and the distillate received into dilute hydrochloric acid. About three-fourths is distilled over. The distillate is then evaporated to dryness with excess of pure platinic chloride (free from nitro-hydrochloric acid). The dry residue is now treated with a mixture of two volumes of absolute alcohol and one of ether, collected on a weighed filter, washed with the said ether mixture, dried at 212° F., and weighed. The following calculation is now applied:—

$$\frac{\text{Weight of the double chloride} \times 36}{442} = \left\{ \begin{array}{l} \text{ammonium in the quantity} \\ \text{taken for analysis.} \end{array} \right.$$

DIVISION III. GRAVIMETRIC ESTIMATION OF ACIDULOUS RADICALS.

1. CHLORIDES.

As Argentic Chloride.

(Practise upon ·5 gramme pure NaCl.)

The solution containing the chloride is precipitated with argentic nitrate. Nitric acid is then added, and the whole warmed and stirred till the liquid is perfectly clear. The precipitate is now treated as directed (see Silver, p. 132). After weighing the chloride it is calculated to Cl.

2. IODIDES.

3. BROMIDES.

4. CYANIDES.

The process for each of these is practically the same as for chloride. The argentic cyanide is, however, collected and weighed upon a weighed filter. The argentic iodide and bromide are treated like the chloride; but *if* a filter is used, it must be a weighed one. The filter is afterwards reweighed, and the increase in weight is the amount of argentic iodide or bromide lost during the washing by decantation.

5. ESTIMATION OF AN IODIDE IN THE PRESENCE OF A CHLORIDE AND A BROMIDE.

By Palladium.

The solution, slightly acidified, is precipitated with excess of palladious chloride or nitrate. The whole is then allowed to stand in a warm place for twenty-four hours, so that the precipitate may thoroughly settle. The supernatant liquor is poured off, and the precipitate having been collected on a filter, and washed, is placed in a weighed platinum crucible and ignited. The whole is then again weighed, and the weight, less that of the crucible, equals the amount of metallic palladium left after ignition. Now, 100 parts of the metal combine with 239·6 parts of iodine in the precipitate, therefore the weight of the metal left after ignition $\times$ 2·396 = the amount of iodine in the weight of the sample taken for analysis.

6. MUTUAL ESTIMATION OF CHLORIDE, BROMIDE, AND IODIDE IN THE PRESENCE OF EACH OTHER.

(Practise upon ·2 gramme each of KCl, KI, and KBr dissolved together in 150 c.c. of water.)

Observation has revealed that if argentic chloride be digested with potassium bromide it is thereby converted into argentic bromide; again, if argentic bromide be digested with potassium iodide it is in turn changed to argentic iodide.

This being recognised, and a solution of a known quantity of the mixed salts being made, divided into three equal volumes, and slight excess of argentic nitrate added to each, three precipitates are obtained, which are to be washed till free from soluble matter. The first of these must then be dried and weighed; the second digested with solution of KBr, washed, dried, and weighed; the third digested with solution of KI, washed, and also, after being rendered free from moisture, weighed. From these three results the amounts of Cl, Br, and I are to be deduced by a calculation based on the difference of the molecular weight of AgCl, AgBr, and AgI.

The first weighing is—

$$AgCl + AgBr + AgI.$$

The second is—

$AgBr + AgI$, and is consequently increased as 35·5 is to 80.

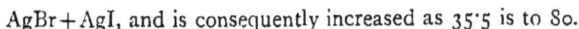

The third is entirely AgI, and is consequently increased as 80 is to 127.

7. SULPHIDES.

By Conversion into Sulphate.

(Practise upon ·5 gramme of purified "black antimony.")

Are analysed by fusion with a large excess of a mixture of potassium nitrate and carbonate, extracting the fused mass with water, filtering, acidulating with hydrochloric acid, adding excess of barium chloride, and proceeding as for a sulphate; but calculating at the last to sulphur instead of sulphuric acid. Some sulphides can be dissolved in nitric acid with the addition of successive small crystals of potassium chlorate. Excess of hydrochloric acid is then added, and the whole evaporated to dryness. The mass is then boiled with dilute hydrochloric acid, filtered, and the filtrate precipitated with barium chloride as above.

8. SULPHATES.

As Barium Sulphate.

(Practise upon ·5 gramme of $MgSO_4 \cdot 7H_2O$.)

To the solution of the sulphate hydrochloric acid is added, then excess of barium chloride, and the whole boiled. When quite clear, a little more barium chloride is added, to ascertain whether all the sulphuric acid has precipitated. The precipitate is now treated precisely as in the barium estimation.

9. NITRATES.

(*A*) In Alkaline Nitrates.

If nitric acid be required to be estimated in, say, ordinary nitre, the sample must first be heated to fusion to remove moisture, and then be quickly powdered. A known quantity of it is now mixed in a platinum crucible with (exactly) four times its weight of plumbic sulphate. The mixture is ignited till it ceases to lose weight, and the loss will just represent the amount of nitric anhydride in the sample taken for analysis.

If plumbic sulphate is used, the reaction is represented by the following formula:—

$$PbSO_4 + 2KNO_3 = PbO + K_2SO_4 + 2NO_2 + O.$$

(*B*) By Conversion into Nitric Oxide.

Already described at page 123.

(*C*) By Conversion into Ammonia.

The nitrate is converted into ammonia by the action of nascent hydrogen, thus—

$$HNO_3 + 4H_2 = NH_3 + 3H_2O.$$

The nascent hydrogen may be applied in various ways, as follows :—

1. By distillation with sodium hydrate and metallic aluminium, and receiving the evolved ammonia into a known volume of standard acid.
2. By acting on the nitrate for 12 hours with zinc or iron and dilute sulphuric acid, and then adding excess of sodium hydrate, and distilling off the ammonia into a known volume of standard acid.

In any case, the standard acid used is then volumetrically checked back by standard sodium hydrate, and the excess of acid used over that of alkali gives the amount of standard acid neutralised by the ammonia. This amount, having been multiplied by the strength of the acid in each c.c., is calculated to ammonia. Then—

$$\frac{\text{Ammonia found}}{17} \times 54 = \left\{ \begin{array}{l} \text{amount of anhydrous nitric acid in} \\ \quad \text{the quantity taken for analysis.} \end{array} \right.$$

10. PHOSPHATES.

(*A*) Estimation of the Strength of Free Phosphoric Acid.

7·38 grammes of the acid are evaporated in a weighed dish with 18·0 grammes of pure litharge, and the dry residue ignited and its weight noted. Thus treated, strong B.P. acid should yield 21·5 grammes of residue.

(*B*) As Magnesium Pyrophosphate in Alkaline Phosphates.

(Practise upon ·75 gramme of pure $Na_2HPO_4 \cdot 12H_2O$.)

They are at once precipitated with ammonium hydrate and *magnesia mixture*, and the precipitate is treated as directed under Magnesium. Should they contain the acid as meta- or pyro- acid, they must first either be boiled with strong nitric acid for one hour, or be fused with potassium sodium carbonate.

(*C*) As Magnesium Pyrophosphate in the Presence of Lime and Magnesia.

(Practise upon ·5 gramme of pure $Ca_3(PO_4)_2$ dissolved in dilute HCl.)

The solution (which must contain orthophosphoric acid, or, failing that, should be boiled with HNO_3 as above) is precipitated with ammonium hydrate, the precipitate redissolved in the smallest amount of acetic acid, the lime precipitated with ammonium oxalate, and the phosphoric acid precipitated in the filtrate by adding ammonium hydrate and *magnesia mixture*, and proceeding as already described for magnesium. Before precipitation this filtrate should be evaporated to a bulk of 3 ounces. This process is suitable for determining the "*soluble phosphates*" in an artificial manure.

(*D*) As Magnesium Pyrophosphate in the Presence of Iron and Aluminium.

(Practise upon ·75 gramme of B. P. *Ferri phosphas* dissolved in dilute HCl.)

The solution is mixed with excess of ammonium acetate, boiled, and ferric chloride added till a dark-brown precipitate forms. This is washed with boiling water and redissolved in a small quantity of dilute HCl. About five grammes (or more) of citric acid are now added, and ammonium hydrate is added to the whole in large excess, and, after cooling, *magnesia mixture* and the precipitate treated as for magnesium already described. If the addition of excess of NH_4HO does not produce a clear lemon-yellow solution then enough citric acid has not been added.

(*E*) Estimation as Phosphomolybdate.

If necessary, the acid solution is heated and precipitated with H_2S to remove arsenic. The excess of H_2S is boiled off, and large excess of nitric acid is added. An excess of ammonium molybdate and ammonium nitrate dissolved in nitric acid is now poured in, the liquid boiled, and finally allowed to stand for some hours in a warm place. The precipitate is filtered off, washed with dilute alcohol until free from acidity, redissolved in dilute ammonium hydrate, and this solution evaporated in a weighed dish on the water bath and dried in the water oven. Its weight $\div$ 28·5 $= P_2O_5$ present. This process is the best for determining small amounts of total phosphates in cast iron, waters and soils.

11. ESTIMATION OF TOTAL AND SOLUBLE PHOSPHATES IN AN ARTIFICIAL MANURE OR OF TOTAL ONLY IN SOIL.

(*A*) Total Phosphates.

About two grammes of the finely powdered substance are weighed accurately, transferred to a beaker and decomposed with HCl, and where necessary with the addition of a drop or two of HNO_3. The solution is then evaporated to dryness in a water bath, taken up with HCl, and after digestion the insoluble silicious matter is separated by filtration; a weighed quantity of citric acid is added; the solution heated up nearly to boiling point, and a weighed quantity of ammonium oxalate added. The quantities used must vary with the substance under examination, the knowledge only being acquired by experience; but it is seldom necessary to add more than 2 grammes citric acid or 2·5 grammes ammonium oxalate. The free acid is then just neutralised with dilute ammonia, and acetic acid added, to decidedly acid reaction. The liquid is kept simmering for a few minutes with constant stirring, and after standing a short time the calcium oxalate is filtered out. Great care must be observed not to have too large an excess of ammonium oxalate present, as magnesium oxalate in an ammoniacal solution is somewhat easily precipitated. To the filtrate ammonia of ·880 sp. gr. is added to about one-fourth of the bulk; and to the liquid, which must remain clear, or only slowly throw down a small precipitate, due to the magnesia present, *magnesia mixture* is added in moderate excess. The liquid must be set aside with occasional stirring for the precipitate to form—the time required being principally determined by the quantity of alumina present. It is best, however, to allow it to stand overnight, although in cases where the alumina is absent, or small, the precipitation will be found to be complete in two hours. The precipitate is then separated from the liquid by filtration, dissolved in as little HCl as possible, and reprecipitated with one-third of its bulk of ammonia. After allowing to stand for two hours with occasional stirring, it may be filtered, and after drying converted, by ignition in a weighed platinum crucible, into $Mg_2P_2O_7$, and weighed as such.

The calcium oxalate is converted into $CaCO_3$ by gentle ignition, weighed, dissolved in HCl, and tested for P_2O_5, which may be present in small quantities, and which should be determined.

The $Mg_2P_2O_7$ is calculated to $Ca_3(PO_4)_2$ unless a full analysis is being made, when it is calculated to P_2O_5 and divided *pro rata* among the bases actually found to be in combination with it.

(*B*) Soluble Phosphates.

Five grammes of the manure are well triturated in a mortar with distilled water, washed into a stoppered 250 c.c. flask and made up to the mark with water. After standing with occasional shaking for two hours 100 c.c. (= 2 grammes of sample) is drawn off by a pipette into a beaker, 2 grammes of citric acid and 2·5 grammes of ammonium oxalate are dissolved in the liquid, which is then treated with ammonia, acetic acid, etc., as above described. If the amount of soluble calcium comes out low, the process should be repeated, using such a weighed quantity of ammonium oxalate as will just remove it from solution. This is because the great source of error in phosphate estimations is the use of excess of oxalate, causing the precipitation of magnesium oxalate with the magnesium ammonium phosphate. The $Mg_2P_2O_7$ is calculated to $Ca_3(PO_4)_2$ and reported as "phosphate made soluble."

12. ARSENIATES

are estimated precisely like phosphates; but the precipitate of ammonium magnesium arseniate is dried at 220° F. on a weighed filter, as already directed under arsenic. The precipitate thus dried is $2(MgNH_4AsO_4) \cdot H_2O$; or, for simplicity of calculation, $MgNH_4AsO_4 \cdot \frac{1}{2}H_2O$.

13. ARBONAS.

A carbonate is estimated by the loss of weight it undergoes by the displacement of its carbonic anhydride by an acid. A small and light flask is procured, and fitted with a cork through which passes a tube (c) containing

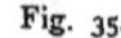

Fig. 35.

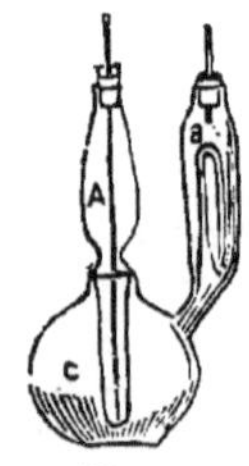

Fig. 36.

fragments of calcium chloride (see fig. 35). A weighed quantity of the carbonate is introduced into the flask with a little water, and a small test-tube (T) about two inches long is filled with sulphuric acid and dropped into the flask, so that, being supported in an upright position, none of the acid shall mix with the carbonate. The cork is put in, and the weight of the whole apparatus having been carefully noted, it is inclined so as to allow the acid to run from the small tube into the body of the flask. Effervescence sets in, the carbonate is dissolved, and the CO_2, escaping through the calcium chloride tube, is deprived of any moisture it might carry with it. When all action has ceased, and the whole has cooled, it is once more weighed. The difference between the two weights gives the amount of CO_2 evolved.

A better apparatus is that figured (No. 36), in which (C) is the flask, (A) the tube to contain HCl to decompose the carbonate, and (B) a tube containing strong H_2SO_4 through which the evolved CO_2 must pass, and so be perfectly deprived of moisture.

14. OXALIC ACID.

As Calcium Carbonate.

The solution containing the acid, or its potassium or sodium salt, is made alkaline with ammonium hydrate, and precipitated with. calcium chloride. The precipitate is washed till free from *chlorides*, dried, ignited, and finally weighed as carbonate. (See *Estimation of Calcium.*)

15. TARTARIC ACID.

As Calcium Oxide.

The solution (which must contain no other bases than Ka, Na, or NH_4) is made *faintly* alkaline by ammonium hydrate, and precipitated by excess of calcic chloride. The precipitate is washed, dried, ignited (with the blowpipe), and weighed as calcic oxide.

16. SILICIC ACID.

(*A*) In oluble Silicates.

By soluble silicates are meant those which are either soluble in water or in hydrochloric acid. The solution (which must contain some free HCl) is evaporated to dryness on the water bath, and the residue dried for an hour at 248° F. After cooling, the mass is moistened with strong hydrochloric acid, and then boiled with water, thus leaving an insoluble residue of pure silica—SiO_2—which is collected on a filter, washed, dried, ignited, and weighed.

(*B*) In Insoluble Silicates.

These bodies must be decomposed by mixing a weighed quantity of the finely powdered substance with four times its weight of sodium potassium carbonate, and fusing the whole for about half an hour. (*When alkalies have to be estimated, a separate special fusion must be made with barium hydrate, or pure lime mixed with ammonium chloride, instead of the double carbonate.*) The crucible must be well covered during fusion. After cooling, the residue is treated with dilute and warm HCl until effervescence ceases, evaporated to dryness, and treated as above described at 248° F., etc.

DIVISION IV. QUANTITATIVE SEPARATIONS.

This department is beyond the scope of the present edition. When the student has practised all the contents of the book up to this point, he will already have a sufficiently general idea of chemical analysis to enable him to fix the line of work he desires to make his speciality. If this be mineral analysis, he must pass to a larger book such as "Fresenius" to complete his knowledge. So as to give, however, some idea of how the preceding processes may be joined in performing the full analysis of a mixture, we take the following example, because it is a standard one, and give a sketch of the manner of working in performing—

The Full Analysis of the Mineral Contents of a Sample of Ordinary Potable Water.

Step I. Take the total solid residue of 100 c.c. (calculated in grains per gallon) as directed in Chapter X., to serve as a check on the results; then ignite and again weigh : loss=*organic and volatile matter.*

Step II. Evaporate 2000 c.c. of the water to dryness in a large porcelain dish, and heat gently till fumes of NH_3 cease to be evolved. Moisten the residue with 10 c.c. of distilled water, and then add 200 c.c. of proof spirit; having gently detached it all from the dish, filter and wash with proof spirit till practically nothing more dissolves. This procedure is useful because it separates the salts present thus:—

(*a*) The filtrate may contain all salts of K and Na, chlorides and nitrates of Ca and Mg, and the sulphate of Mg.

(*b*) The insoluble residue may contain the sulphate and carbonate of Ca and the carbonate of Mg, together with any iron silicous matter present.

(1) *Analysis of the filtrate.*

(*a*) Evaporate till the spirit is driven off, cool, transfer to a 250 c.c. flask, and make up to the mark with distilled water. Divide into two portions of 100 c.c. and 150 c.c. respectively, marking them (A) and (B).

(*b*) To (A) add NH_4Cl, NH_4HO and $(NH_4)_2C_2O_4$, to precipitate the Ca, and estimate as usual as $CaCO_3$. Calculate to CaO and then × 2·5=CaO present as Cl or NO_3 in the original 2000 c.c. of water taken.

(*c*) To filtrate and washings from (*b*), concentrated to 50 c.c. and cooled, add Na_2HPO_4 to precipitate Mg, treat as usual, and weigh as $Mg_2P_2O_7$. Calculate to MgO, and then × 2·5=MgO present as Cl or NO_3 in the original 2000 c.c. of water.

(*d*) To (B) acidulate with HCl and add $BaCl_2$ to precipitate sulphate. Treat as usual and weigh as $BaSO_4$. Calculate to SO_3 and then × 2·5 ÷ 1·5 = total SO_3 present in the original 2000 c.c. of water taken in combination with K or Na.

(*e*) The filtrate and washings from (*d*) are evaporated to a low bulk rendered alkaline with pure $Ca(HO)_2$, the separation for alkalies given at page 140 gone through, and the K and Na present both estimated as chlorides. Results × 2·5 ÷ 1·5 = total K and Na present in the 2000 c.c. of water started with.

(*f*) The residue from (*e*) is dissolved in a little water and the K estimated thereon by $PtCl_4$ in the usual manner and calculated to K_2O (see page 139). An equivalent amount of KCl (calculated from this K_2O) is then deducted from residue (*e*) and the balance is NaCl, which is calculated to Na_2O.

(2) *Analysis of the insoluble portion.*

(*a*) This is washed from the filter with distilled water and then boiled with 100 c.c. of H_2O and HCl added till effervescence ceases.

Any insoluble is filtered out, washed with boiling H_2O, dried, ignited and weighed=*silicious matter* in the 2000 c.c. of water started with.

(*b*) The filtrate and washings are warmed with a drop or two of HNO_3 and mixed with $NH_4Cl + NH_4HO$, and the iron estimated if present as Fe_2O_3, and result calculated to Fe=total Fe in the 2000 c.c. of water taken.

(*c*) Divide filtrate and washings into two equal parts, A and B.

(*d*) The portion A is precipitated with $(NH_4)_2C_2O_4$, and the calcium estimated as $CaCO_3$ and calculated to CaO=total CaO as carbonate or sulphate in the original 2000 c.c. of water taken.

(*e*) The filtrate from (*d*) is concentrated to a low bulk, cooled, and Na_2HPO_4 added with excess of NH_4HO, and the Mg estimated as usual as $Mg_2P_2O_7$ and calculated to MgO. Result × 2= total MgO present as carbonate or sulphate in the original 2000 c.c. of water.

(*f*) The portion B is acidulated with HCl, and the sulphate estimated by $BaCl_2$, weighed as $BaSO_4$ and calculated to SO_3. Result × 2=total SO_3 in combination with Ca or Mg in the original 2000 c.c. of water.

Step III. Evaporate 250 c.c. of the water to a bulk of 2 c.c. and treat in the nitrometer to estimate the nitric acid (see page 123). Resulting NO calculated to N_2O_5 and × 8= total N_2O_5 present in 2000 c.c. of water.

Step IV. Take the amount of chlorides volumetrically (page 113) in 100 c.c. of the water. Result × 20=Cl in 2000 c.c. of water.

Step V. *Calculation of results.*

(*a*) All our results being in grammes or fractions of the same from 2000 c.c. of water, each must be multiplied by 35, which will bring them all to grains per gallon (parts in 70,000). Now we state out our analysis thus (example taken from actual practice):—

A sample of water yielding 20·1 grains of total solids per gallon, of which ·45 grain was "organic and volatile matter," and the balance (19·65 grains) was mineral matter, showed on analysis :—

	grains per gallon
(*a*) In the portion soluble in spirit :	
Potassium oxide	·2704
Sodium oxide	1·4097
Chlorine	1·2133
Sulphuric anhydride	·6803
Calcium oxide	none.
Magnesium oxide	none.
Nitric anhydride	none.
(*b*) In the portion insoluble in spirit :	
Calcium oxide	7·3953
Magnesium oxide	1·0000
Sulphuric anhydride	1·7647
Silicious matter	·2000
Ferric oxide	·0500
Total found	13·9837

From this residue is now to be always deducted an amount of oxygen equivalent to the chlorine found, because all the bases have been calculated to oxides, while haloid salts contain no oxygen. The chlorine found is 1·2133, and—

$$\frac{1·2133 \times 16}{71} = ·2737 \text{ oxygen, equivalent to Cl found.}$$

Performing this deduction, we have :

13·9837 − ·2737 = 13·7100 grains of solid matter actually found.

The total residue, after driving off organic matter, was 19·65 grains per gallon, and the difference is due to CO_2 unestimated, thus :—

19·65 — 13·71 = 5·94 grains of CO_2 per gallon.

Adding now this CO_2 to the substances actually estimated, we get :

Total substances found + CO_2 = 19·65,
Actual residue found = 19·65,

which proves our analysis to be correct.

It now remains to calculate how these bases and acids are probably combined as salts actually present, by the following general rules of affinity, thus :—

(*a*) *In the portion soluble in spirit.* (1) Any sulphuric anhydride will prefer the bases in the following order : K, Na, Ca, Mg. (2) Chlorine will prefer the bases in the same order *after the SO_3 is satisfied.* Therefore we first calculate our K_2O to K_2SO_4, which gives ·50 and uses up ·2296 of our SO_3, and the balance of SO_3 (·4507) we calculate to Na_2SO_4. This gives ·80 Na_2SO_4, and leaves 1·0604 Na_2O not as sulphate and therefore existing as chloride. Calculating accordingly, we get 2·00 of NaCl, which just uses up all our chlorine. Therefore this portion contained altogether :—

Potassium sulphate	·50
Sodium sulphate	·80
Sodium chloride	2·00

(*b*) *In the portion insoluble in spirit* : (1) The SO_3 found will all be present as $CaSO_4$, and the balance of CaO and all the MgO will be as carbonates. Therefore 1·7647 SO_3 calculated to $CaSO_4$ becomes 3·00 and uses up 1·2353 of CaO, leaving 6·16 to be calculated to $CaCO_3$. This yields 11·00 $CaCO_3$, and the 1·00 of MgO found calculated to $MgCO_3$ gives 2·10. Thus this portion contains :

Calcium sulphate	3·00
Calcium carbonate	11·00
Magnesium carbonate	2·10

Putting now the whole analysis together, we have

Potassium sulphate	·50
Sodium sulphate	·80
Sodium chloride	2·00
Calcium sulphate	3·00
Calcium carbonate	11·00
Magnesium carbonate	2·10
Ferric oxide	·05
Silica	·20
Organic and volatile matter	·45
Total residue	20·10.

CHAPTER IX.

ULTIMATE ORGANIC ANALYSIS.

THIS process consists in estimating the amount of each element present in any organic compound, as distinguished from proximate analysis, which estimates the amounts of the compounds themselves.

I. LIST OF APPARATUS REQUIRED.

1. A combustion furnace, which is a series of Bunsen burners arranged in a frame so as to gradually raise a tube placed over them to a red heat. The tube lies in a bed made of a series of fire-bricks, which confine the heat and make it play all round the tube (see fig. 37).
2. Combustion tubes made of hard glass, not softening at a red heat. These tubes are closed at one end by drawing out before the

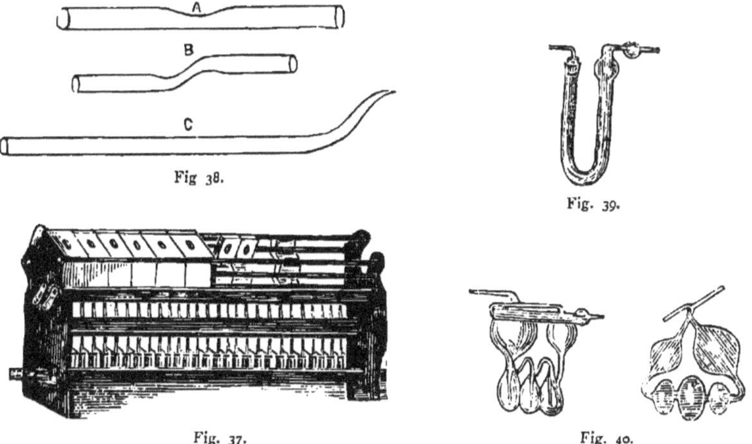

Fig 38.

Fig. 39.

Fig. 37.

Fig. 40.

blowpipe and turning up. The mode of doing this is illustrated in fig. 38.
3. U tubes packed with perfectly dry calcium chloride in small fragments, so as to allow the free passage of gases (illustrated in fig. 39), and hereafter referred to for brevity as "$CaCl_2$ tubes."
4. Bulbs charged with strong solution of potassium hydrate (1 in 1), hereafter for brevity referred to as "KHO bulbs." Two common forms of such bulbs are illustrated in fig. 40.

5. Bulbs for absorbing ammonia and intended to be charged with dilute acid of known strength, and hereafter referred to as "nitrogen bulbs" for brevity. Two common forms of such bulbs are illustrated in fig. 41.
6. Graduated tubes closed at one end to hold 50 c.c., and graduated from the closed end downwards in $\frac{1}{10}$ths of a c.c., used for collecting gases and measuring them after collection, hereafter referred to as "gas collecting tubes" for brevity (fig. 42).
7. A deep cylindrical vessel of glass filled with water and furnished with a thermometer dipping in the water. The whole sufficiently high to permit of the entire immersion of the gas measuring tubes, and wide enough at the top to admit the hand, as shown in fig. 43. This is hereafter called the "measuring trough" for brevity.
8. Glass towers, filled in one limb with fragments of $CaCl_2$ to absorb moisture, and in the other with fragments of soda lime to

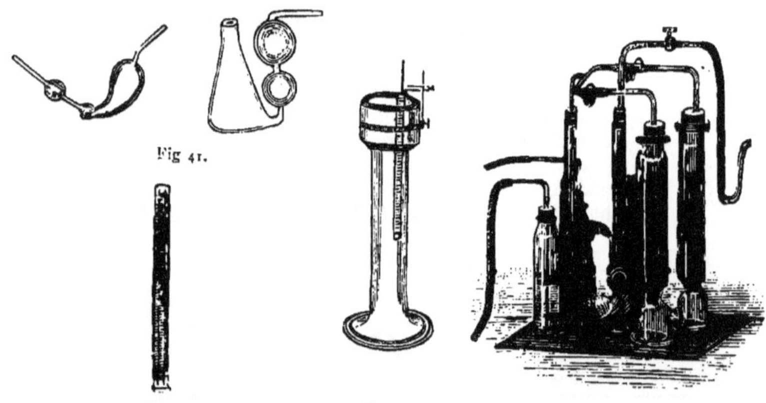

Fig 41.

Fig 42. Fig 43. Fig 44.

absorb carbon dioxide. These are illustrated in fig. 44, and are used for freeing any air which may be caused to pass through them from moisture and CO_2; hereafter called "air purifying towers" for brevity.

II. ESTIMATION OF CARBON AND HYDROGEN.

This process is performed by heating a weighed quantity of the substance in a tube with some body readily parting with oxygen at a red heat, such as cupric oxide or plumbic chromate, by which the hydrogen and carbon of the organic body are respectively oxidised into water and carbonic anhydride. These products are passed first through a previously weighed tube containing calcium chloride, which retains the water, and then through a bulb apparatus containing potassium hydrate, which absorbs the carbonic anhydride and has also been previously weighed. After the experiment is finished, the increase in weight of the tubes is calculated thus:—

$$\text{As } H_2O : H_2 : : \text{increase in } CaCl_2 \text{ tube} : x.$$
$$\text{As } CO_2 : C : : \text{increase in KHO bulbs} : x.$$

The details of the actual process are as follows:—

Cupric oxide is prepared by heating cupric nitrate to bright redness in a crucible; it is reduced to powder while still warm, and preserved in a well-stoppered bottle. A combustion tube is procured, having a bore of about half an inch and a length of fifteen to eighteen inches. Sufficient cupric oxide to fill it is measured out, heated in a hard glass flask to expel any moisture (as it is very hygroscopic), well corked, and allowed to cool. About 4 decigrammes of gently fused and cooled potassium chlorate is first dropped into the combustion tube, and then when cold a little oxide is introduced. About ·4 gramme of the organic substance having been weighed out, it is rubbed up in a warm mortar with enough cupric oxide to half fill the tube, and the mixture is quickly transferred to that apparatus; sufficient cupric oxide to nearly fill the tube is then introduced, and lastly a loose plug of asbestos is inserted to keep all in place. To this charged tube is then attached by means of a good cork a previously weighed "$CaCl_2$ tube," and to this is in turn fixed by an indiarubber joint a set of "KHO bulbs" also previously weighed. The tube is now placed in the furnace, the glass attachments supported outside of it by appropriate stands, and made perfectly air-tight. Heat is now applied to the front portion of the tube by lighting the first five or six burners, and when that part is quite red-hot the next burner is turned on. The heat is thus applied gradually, taking care that it is so regulated as to produce a regular slow passage of the evolved gases, so that the bubbles may be distinctly counted as they pass through the "KHO bulbs." When the whole tube is heated from end to end the potassium chlorate at the back gives off a little oxygen, which sweeps the last traces of moisture and carbonic anhydride into the bulbs, which are then detached, weighed, and the increase in weight of each noted.

Weight of sugar taken	·475 gramme.
Potash bulbs after combustion weighed	79·113 grammes.
,, ,, before ,, ,,	78·382 ,,
Difference, due to CO_2	·731 ,,
Calcium chloride tube after combustion weighed	23·605 grammes.
,, ,, before ,, ,,	23·330 ,,
Difference, due to H_2O	·275 ,,

$\frac{(C)\ 12 \times \cdot 731}{(CO_2)\ 44} = \cdot 1994$ carbon $\qquad \frac{(H_2)\ 2 \times \cdot 275}{(H_2O)\ 18} = \cdot 03056$ hydrogen.

Total sugar taken	·475
Total C and H found	·22996
Difference, due to oxygen	·24504

Or, in percentage—	
Carbon	41·98
Hydrogen	6·43
Oxygen	51·59
	100·00

When the organic matter is a liquid, it is introduced into a small drawn-out tube previously tared; the end is hermetically sealed by the blowpipe and the whole again weighed, thus obtaining the weight of the liquid. A little oxide having been first put into the combustion tube, the sealed one is dropped in, its end broken by a wire, and the whole of the rest of the oxide poured in. The heat is applied till six or seven inches of CuO are bright red, and then the burner underneath the spot where the tube with the liquid lies is cautiously turned on, so as to volatilise the vapor and cause it to pass over the red-hot cupric oxide and so suffer combustion. Fats and other bodies which cannot be mixed with the oxide are weighed in a small platinum boat, which is dropped in and treated like the tube of liquid already referred to.

Special Notes to the Foregoing Process.

(*a*) Substances containing sulphur, phosphorus, arsenic, chlorine, or any halogen, are best mixed with fused and powdered plumbic chromate instead of cupric oxide, so as to avoid the formation of volatile cupric compounds.

(*b*) When the substance also contains nitrogen, the front part of the combustion tube must be plugged with a roll of bright copper gauze about 4 inches long instead of the asbestos. This is to reduce any oxide of nitrogen, which if allowed to pass into the potash would be absorbed and count as carbon dioxide. The copper, however, takes the oxygen, and only leaves nitrogen, which passes unabsorbed.

(*c*) Very refractory bodies should be burned in a current of air or of oxygen carefully deprived of any moisture or carbon dioxide by causing it to pass through "purifying towers" (see page 151) before entering the tube.

III. ESTIMATION OF NITROGEN.

The **estimation of nitrogen** in all compounds, not being nitrites or nitrates, is conducted as follows:—

(*a*) **The Method of Varrentrapp and Will.** This depends for its success on the fact that when nitrogenous substances are strongly heated with **sodium hydrate,** they are decomposed, forming a carbonate and oxide with the oxygen from the hydroxyl, and liberating hydrogen, which then combines with the nitrogen to form **ammonia.**

So as to prevent fusion of the glass tubes employed, solution of sodium hydrate is evaporated to dryness with calcium oxide, and the resulting mixture, known as **soda-lime,** is heated to redness and preserved for use.

About ·5 gramme of the perfectly dry nitrogenous substance is mixed in a warm wedgwood mortar, with sufficient soda-lime to fill about three-quarters of a combustion tube. The mortar is rinsed with a little more soda-lime, and the rinsings also put into the combustion tube. More soda-lime is then put into the tube, until it is filled to within two inches of its open end, and then an inch of asbestos in shreds packed loosely in front of all, so as to prevent the passage of fragments of lime along with the evolved gases. A set of "nitrogen bulbs," containing a little tolerably strong hydrochloric acid, is adapted to the mouth of the combustion tube by means of a well-fitting cork, and the tube is placed in the combustion furnace, so that the bulbs and about an inch of the tube project outside. A few front burners having been lighted, and the free part of the tube having become red-hot, the heat is gradually applied until the whole of it has been heated to bright redness. When this point is attained, and bubbles of gas cease to pass through the acid in the bulbs, the end of the tube is broken off, and some air sucked through the apparatus by means of a small tube attached to the outer end of the bulbs. The contents of the bulbs are then transferred into a small basin, the bulbs washed out into it, first by means of a little alcohol, and afterwards repeatedly with distilled water. An excess of platinic chloride is then added, the whole evaporated to dryness in a water bath, and the precipitate, having been moistened with a little alcohol, is washed on to a weighed filter. The washing with alcohol is continued until it passes colorless, and the precipitate is dried at 212° and weighed. The weight of the filter having been deducted, the balance is $PtCl_4(NH_4Cl)_2$, which is then calculated to N_2.

A more simple plan, adapted to the estimation of **nitrogen in manures and other commercial products**, may be thus followed out:—

The ammonia is received into 20 c.c. of semi-normal volumetric solution of oxalic or sulphuric acid, which is placed in the bulb apparatus. The ammonia which passes into the bulbs during the combustion neutralises a portion of this acid, and at the conclusion of the operation the amount of acid still remaining free is ascertained by means of a corresponding normal volumetric solution of sodium hydrate (see VOLUMETRIC ANALYSIS, p. 112). The difference between the amount of acid originally placed in the bulbs and that remaining free, as thus ascertained, evidently corresponds to the amount of sulphuric acid neutralised by the ammonia produced during the combustion. The strength of the acid being known, a simple calculation enables us to ascertain the amount of the nitrogen evolved in the form of ammonia.

(*b*) **The Process of Dumas.** This consists in measuring the amount of pure nitrogen evolved, and is suitable for certain organic bases and for compounds containing nitrosyl (NO) or nitryl (NO_2), in which the soda-lime fails to convert all the nitrogen into ammonia.

The combustion tube (which in this case is twenty-six to twenty-eight inches long) is packed (1) with six inches of dry sodium hydrogen carbonate; (2) with a little pure cupric oxide; (3) with the weighed substance mixed with CuO; (4) with more pure CuO; and lastly with a considerable length of pure spongy metallic copper; and the whole is closed by a good cork through which passes a bent delivery tube dipping under the surface of mercury in a small pneumatic trough. Heat is first applied to the very end portion of the $NaHCO_3$ until sufficient CO_2 has been given off to entirely drive all the air out of the apparatus, which is ascertained by collecting a little of the gas passing off and seeing that it is entirely absorbed by solution of potassium hydrate. A graduated "gas collecting tube" is then filled, one-third with strong solution of KHO, and the remainder with mercury, and carefully inverted into the mercury trough so that no air is admitted, and placed over the mouth of the delivery tube. Combustion is now commenced at the front of the tube and gradually carried backwards as usual. The gases evolved are CO_2 and N, the former of which is absorbed by the KHO and the latter collects in the graduated jar. When the heat reaches the back of the tube the remainder of the $NaHCO_3$ is decomposed, and the carbonic anhydride given off chases any trace of nitrogen out of the tube. The collecting tube is then closed by a small cup containing mercury, and transferred to the "measuring trough," and entirely immersed therein. After leaving it until its contents have acquired the temperature of the water, it is raised so that the level of the water inside and outside the tube is equal, and the volume is read off. The temperature and pressure being noted, the weight of the nitrogen is obtained by the species of calculation already described at page 101.

(*c*) **Kjeldahl's Process.**—This method is rapidly superseding combustion, being almost universally employed in Germany. It is as follows:—

The substance, whether solid or liquid—and, if solid, it need not even be powdered—is weighed or measured into a small flask, in which the subsequent treatment is executed. In the case of substances containing about 10 per cent. of nitrogen, about 0·1 to 0·2 grm. are sufficient; if less is present, up to 1 grm. may be taken. Ten c.c. of strong sulphuric acid, known to be free from ammonia, are now introduced by means of a pipette, the flask, in an inclined position, placed upon wire gauze over a gas flame, and the contents heated for one or two hours to near the boiling point of the acid, which may be known from the occasional slight bumping. If the substances retain their nitrogen very tenaciously, such as quinine, morphine, etc., a little fuming sulphuric or anhydrous phosphoric acid may be added to the contents of the flask in the beginning. When the action of the acid is completed, which usually results in a solution of the solid, the flame is removed, and a fine spray of finely powdered permanganate of potassium, contained in a tube closed with a fine wire gauze and inserted in the neck of the flask, is made to rain down upon the acid in the flask until the latter assumes a green color. This step is only, however, necessary when the substance is specially difficult to oxidise, most ordinary bodies being sufficiently attacked by fuming sulphuric acid. The flask is then cooled, the contents diluted with water, 40 c.c. of a solution containing 35 per cent. of rochelle salt and 30 per cent. of sodium hydrate are added, and the flask immediately connected with and its contents distilled into a well-cooled condenser, provided with a receiver containing standard volumetric acid. If the original flask is too small, the contents may be transferred to another, water being used in portions to wash the original flask carefully. The estimation of the ammonia, by finding the amount of unsaturated acid, is conducted by volumetric soda, as already described in the soda-lime combustion. The ammonia found is calculated to nitrogen. Any substance likely to give off nitrous fumes should be first mixed with twice its weight of sugar. Nitrates require dilution with three times their bulk of benzoic acid. The rochelle salt takes no part in the reaction, but is a device to prevent bumping.

IV. ESTIMATION OF CHLORINE.

Chlorine is estimated by combustion of the substance in a tube filled with pure calcium oxide, when it displaces oxygen and turns part of the oxide into soluble calcium chloride. After combustion the contents of the tube are digested in water, filtered, excess of argentic nitrate added, and the precipitate washed, dried, and weighed, as already directed (see Gravimetric Estimation of Chlorine, p. 141).

V. ESTIMATION OF SULPHUR, PHOSPHORUS AND ARSENIC.

Sulphur, phosphorus and arsenic, are estimated by fusing in a silver crucible, with twelve times the bulk of pure potassium nitrate, a little mixed carbonate and nitrate being first put into the crucible, heated to fusion, and the mixture added a little at a time, waiting between the additions till action ceases. After fusion the sulphur and phosphorus, which have been converted into sulphuric and phosphoric acids respectively, are estimated by dissolving the residue in water and proceeding by precipitation as barium sulphate and magnesium ammonium phosphate respectively. (See Gravimetric Estimation of Sulphuric and Phosphoric Acids.)

CHAPTER X.

THE ANALYSIS OF WATER, AIR, AND FOOD.

DIVISION I. THE SANITARY ANALYSIS OF WATER.

In the following instructions, the lines laid down by the Water Committee of the Society of Public Analysts (of which the author was a member) have been followed.

1. Collection of the Sample.

This is to be taken in a clean stoppered "Winchester quart" bottle, previously entirely filled with the water, and emptied before finally filling up and introducing the stopper. The sample should be kept dark, and analysed with as little delay as possible.

2. Color.

This is to be taken by looking at a column of water in a colorless glass tube 2 feet long, and held over white paper. The presence of a greenish-yellow color is an adverse indication.

3. Odor.

An 8-ounce wide-mouthed stoppered bottle, free from odor, is half filled with the water, warmed in the water bath to 100° Fahr., shaken, and then the stopper is removed and the odor instantly noted. Peaty waters, and those containing marked amounts of sewage, can frequently be thus detected by a practised nose.

4. Total Solid Residue.

Heat a platinum basin of about 130 c.c. capacity to redness, cool it under the desiccator and weigh. Introduce 100 c.c. of the water, and evaporate over a low gas flame until reduced to about 10 c.c.; then place it on the water bath till dry. Finally, heat it in the air bath at 220° Fahr. until it ceases to lose weight, cool under the desiccator, and weigh. Having deducted the tare of the basin, the difference in milligrammes × 10 = total residue in parts per million, or the same × 7 ÷ 10 = grains per gallon. Example:—

Weight dish + residue . .	89·336
Tare of dish	89·300
	·036 or 36 milligrammes;

then $\frac{36 \times 7}{10} = 25·2$ grains per gallon.

The residue should now be gradually heated to redness, and the presence of organic matter carefully looked for as indicated by charring; also the nature of the same, by whether the odor on burning is purely carbonaceous (like burning sugar) or nitrogenous (like burning hair). This latter is an especially unfavorable indication.

5. Chlorine.

Solutions required:—

(*a*) 4·789 grammes *pure* crystallised argentic nitrate, dissolved in 1000 c.c. of distilled water. Each c.c. of this solution=·001 (*one milligramme*) of chlorine.

(*b*) 5 grammes potassium chromate dissolved in 100 c.c. distilled water, and a weak solution of argentic nitrate dropped in until a slight permanent red precipitate is produced, which is allowed to settle in the bottle.

Process.—Put 100 c.c. of the water into a white basin, add a few drops of the chromate solution, and titrate with the silver solution from a burette, graduated in $\frac{1}{10}$ of c.c., until a faint permanent change of color is produced, as already described (Chap. VII. p. 113). Note the number of c.c. used, multiply by 10, and the result will be chlorine in milligrammes per litre (or parts per million), or multiply by 7 and divide by 10, which will give grains per gallon (or parts per 70,000). The water itself must be perfectly neutral; if acid it must be first shaken with a little pure precipitated chalk.

6. Phosphoric Acid.

Solution required:—

One part pure molybdic acid is dissolved in 4 parts ammonia, sp. gr. ·960. This solution, after filtration, is poured with constant stirring into 15 parts of nitric acid of 1·20 sp. gr. It should be kept in the dark and carefully decanted from any precipitate which may form.

Process.—The ignited total residue, obtained as directed in No. 4, is to be treated with a few drops of nitric acid, and the silica rendered insoluble by evaporation to dryness. The residue is then taken up with a few drops of dilute nitric acid, some water is added, and the solution is filtered through a filter previously washed with dilute nitric acid. The filtrate, which should measure 3 c.c. (or say 50 grains), is mixed with 3 c.c. of molybdic solution, gently warmed, and set aside for fifteen minutes at a temperature of 80° Fahr. The result is reported as "traces," "heavy traces," or "very heavy traces," when a color, turbidity, or definite precipitate are respectively produced, after standing for fifteen miuutes.

7. Nitrogen in Nitrates.

(*a*) *Crum Process.*—This is the best process where a nitrometer is available. 250 c.c. of the water must be evaporated to a small bulk, the chlorine precipitated with saturated solution of argentic sulphate, filtered, and the filtrate concentrated in a basin to 2 c.c. A nitrometer is charged with mercury, and the three-way stopcock closed, both to measuring tube and waste pipe. The concentrated filtrate is poured into the cup at the top of the measuring tube, and the vessel which contained it rinsed with 1 c.c. of water, and the contents added. The stopcock is opened to the measuring tube, and, by lowering the pressure tube, the liquid is sucked out of the cup into the tube. The basin is again rinsed with 5 c.c. of pure strong sulphuric acid, and this is also transferred to the cup and sucked into the measuring tube. The stopcock is once more closed, and 12 c.c. more sulphuric acid put into the cup, and the stopcock opened to the measuring tube until 10 c.c. of acid have passed in. The excess of acid is discharged, and the cup and waste pipe rinsed with water. Any gas which has collected in the measuring tube is expelled by opening the stopcock and raising the pressure tube, taking care no liquid escapes. The stopcock is closed, the measuring tube taken from its clamp and shaken by bringing it slowly to a nearly horizontal position and then suddenly raising it to a vertical one. This shaking is continued until no more gas is given off, the operation being, as a rule, quite complete in fifteen minutes. Now prepare a mixture of one part of water with five parts of sulphuric acid, and let it stand to cool. After an hour, pour enough of this mixture into the pressure tube to equal the length of the column of acidulated water in the working tube, bring the two tubes side by side, raise or lower the pressure tube until the mercury is at the same level in both tubes, and read off the volume of the nitric oxide. This volume expressed in c.c.'s and corrected to normal temperature and pressure gives, when multiplied by ·175, the nitrogen in nitrates, in *grains per gallon*, if 250 c.c. of the water have been used. According to some authorities the precipitation of the chlorides is not necessary.

(*b*) *Copper-Zinc Process.*—This must be carried out as follows:—A wet copper-zinc couple is prepared by taking a piece of clean zinc foil, about 3 in. by 2 in., and immersing it in a solution of copper sulphate, containing about 3 per cent. of the pure crystallised salt. A copious and firmly adherent coating of black copper is speedily deposited upon the surface of the zinc, which must be allowed to remain in the solution until the deposit is thick enough, but not for too long a time, or it will become pulverulent and not adhere firmly to the zinc,—three or four minutes will generally be sufficient.

The zinc coated with copper must then be removed from the solution, which may be bottled for subsequent use, and the couple thoroughly washed first with distilled water, and finally with the water to be analysed, in order that this may replace the adhering distilled water. It is then put into a clean 6 or 8-ounce wide-mouth stoppered glass bottle, and covered with 100 c.c. of the water to be analysed. If the water be very soft a small addition, say one part per 1000, of sodium chloride, will accelerate the reaction. The stopper must then be inserted in the bottle and the water allowed to remain overnight in a warm place. If still greater speed be necessary the temperature may be raised to 90° or 100° F. (32° or 38° C.). With hard water it is preferable to add a small quantity of pure oxalic acid to precipitate the lime and quicken the reaction. When the reduction is complete the fluid contents of the bottle are to be transferred to a retort with 200 c.c. of ammonia-free distilled water, and the retort having been attached to a condenser, the contents are distilled till the distillate which comes over gives no color with "Nessler." The distillate is then "nesslerised," as already described (Chapter VII., page 125), and the number of milligrammes of NH_3 found are calculated to N ($\times$ 14 $\div$ 17), and then the resulting milligrammes of N $\times$ 7 $\div$ 10 = grains per gallon, or $\times$ 10 only = parts per million.

8. Ammonia and Albuminoid Ammonia.

These two indications are successively taken on the same quantity of water. The former is an estimation of the ammonia present in the water in the form of ammonium salts or similar compounds readily decomposed by a weak alkali, while the latter shows the ammonia derived from the decomposition of nitrogenous organic matter under the joint influence of an oxidising agent ($KMnO_4$) and a hydrating agent (KHO), and is therefore a measure of the nitrogenous, and consequently presumably dangerous, organic matters contained in the water under examination.

The solutions and apparatus required are:—

(*a*) *Sodium carbonate.*

A 20 per cent. solution of recently ignited pure sodium carbonate.

(*b*) *Alkaline potassium permanganate solution.*

Dissolve 200 parts of potassium hydrate and 8 parts of pure potassium permanganate in 1200 parts of distilled water, and boil the solution rapidly till concentrated to 1000 parts, cool, and keep in a well stoppered bottle.

(*c*) *Distilled water which is free from Ammonia.*

Distilled water which gives no reaction with Nessler test is pure enough. But, if this is not available, take the purest distilled water procurable, add pure ignited sodium carbonate in the proportion of 1 part per 1000, and boil briskly until at least one-fourth has been evaporated.

(*d*) A 40-ounce stoppered retort with a neck small enough to pass loosely into the internal tube of a Liebig's condenser to the extent of 6 inches (see illustration Chapter I. page 4). The joint between the retort and condenser is made by an ordinary indiarubber ring—such as those used for the tops of umbrellas—which has been previously soaked in a dilute solution of soda or potash—being stretched over the retort tube in such a position that when the retort tube is inserted in the condenser it shall fit fairly tightly within the mouth of the tube about half an inch from the end.

(*e*) All the materials for "Nesslerising" (see Chapter VII. page 125).

The process is as follows:—

(*a*) *For free ammonia.*—First test a little of the water with tincture of cochineal, to see if it shows an alkaline reaction. Put 500 c.c. of distilled water into the retort and distil until 50 c.c. of the distillate gives no color with 1½ c.c. of Nessler, thus rendering the whole apparatus "ammonia-free." Let the whole cool, pour out the distilled water (which may be saved for ammonia-free water), put in 500 c.c. of the water to be analysed, and if it had not an alkaline reaction make it alkaline with a drop or two of the sodium carbonate solution. The distillation should then be commenced, and not less than 100 c.c. distilled over. The receiver should fit closely, but not air-tight, on to the condenser. The distillation should be conducted as rapidly as is compatible with a certainty that no spurting takes place. After 100 c.c. have been distilled over, the receiver should be changed, that containing the distillate being stoppered to preserve it from access of ammoniacal fumes. 100 c.c. measuring flasks make convenient receivers. The distillation must be continued until 50 c.c. more are distilled over; and this second portion of the distillate must be tested with Nessler's reagent to ascertain if it contains any ammonia. If it does not, the distillation for free ammonia may be discontinued, and this last distillate rejected; but, if it does contain any, the distillation must be continued still longer, until a portion of 50 c.c., or say 700 grains, when collected, shows no coloration with the Nessler test. The whole of the distillates must be nesslerised in the usual manner, and the total number of milligrammes of ammonia found having been added together are multiplied by 2, which gives milligrammes per litre (parts per million). This number in turn multiplied by 7 and divided by 100 gives grains per gallon of *free ammonia.*

(*b*) *For albuminoid ammonia.*—As soon as the distillation above referred to has been started, 50 c.c. of the alkaline potassium permanganate solution are placed in a flask with 150 c.c. of distilled water and boiled gently during the whole time that the free ammonia is distilling, adding some ammonia-free water if necessary to prevent too much concentration. The object of this is to insure the entire evolution of any trace of ammonia present in the alkaline permanganate, thus avoiding a check analysis, as usually recommended. When the distillation of the free ammonia is complete, take out the stopper of the retort and pour in this boiled alkaline permanganate by means of a perfectly clean funnel with a long limb. Now replace the stopper and continue the distillation, when the albuminoid ammonia will begin to come over. After 200 c.c. have been distilled change the receiver and take off 50 c.c. at a time, as already described, until the last 50 comes over ammonia-free. Mix the distillates, nesslerise and calculate as for the free ammonia, noting the total result as albuminoid ammonia in parts per million or grains per gallon. Great care must be taken that no ammonia is kept in the room devoted to water analysis, and that all receivers used are first insured to be perfectly ammonia-free by proper rinsing with ammonia-free water and testing with Nessler.

9. Oxygen required to oxidise the Organic Matter.

Solutions required :—

(*a*) *Standard solution of Potassium permanganate.*
Dissolve ·395 parts of pure potassium permanganate in 1000 of water. Each c.c. contains ·0001 gramme available oxygen.

(*b*) *Potassium Iodide solution.*
One part of the pure salt recrystallised from alcohol, dissolved in 10 parts distilled water.

(*c*) *Dilute Sulphuric acid.*
One part by volume of pure sulphuric acid is mixed with three parts by volume of distilled water, and solution of potassium permanganate dropped in until the whole retains a *very faint* pink tint, after warming to 80° F. for four hours.

(*d*) *Sodium hyposulphite—*
One part of crystallised sodium hyposulphite dissolved in 1000 parts of water.

(*e*) *Starch water.*
One part of starch to be intimately mixed with 500 parts of cold water, and the whole briskly boiled for five minutes, and filtered, or allowed to settle.

The Process.—Two separate determinations have to be made: viz., the amount of oxygen absorbed during 15 minutes, and that absorbed during four hours; both are to be made at a temperature of 80° F. It is most convenient to make these determinations in 12-oz. stoppered bottles, which have been rinsed with sulphuric acid and then with water. Put 250 c.c. of the water into each bottle, which must be stoppered and immersed in a water bath until the temperature rises to 80° F. Now add to each bottle 10 c.c. of the dilute sulphuric acid, and then 10 c.c. of the standard potassium permanganate solution. Fifteen minutes after the addition of the potassium permanganate, one of the bottles must be removed from the bath and two or three drops of the solution of potassium iodide added to remove the pink color. After thorough admixture, run from a burette the standard solution of sodium hyposulphite, until the yellow color is nearly destroyed, then add a few drops of starch water, and continue the addition of the hyposulphite until the blue color is just discharged. If the titration has been properly conducted, the addition of one drop of potassium permanganate solution will restore the blue color. At the end of four hours remove the other bottle, add potassium iodide, and titrate with sodium hyposulphite as just described. Should the pink color of the water in the bottle diminish rapidly during the four hours, further measured quantities of the standard solution of potassium permanganate must be added from time to time so as to keep it markedly pink. The hyposulphite solution must be standardised, not only at first, but (since it is liable to change) from time to time in the following way :—To 250 c.c. of pure redistilled water add two or three drops of the solution of potassium iodide, and then 10 c.c. of the standard solution of potassium permanganate. Titrate with the hyposulphite solution as above described. The quantity used will be the amount of hyposulphite solution corresponding to 10 c.c. of the standard potassium permanganate solution, and therefore representing 1 milligramme of oxygen consumed. The difference between the number of c.c. of hyposulphite used in the blank experiment and that used in the titration of the samples of water multiplied by the amount of available oxygen contained in the permanganate added (= 1 milligramme if 10 c.c. have been used), and the product divided by the number of c.c. of hyposulphite corresponding to the latter as found by the check experiment, is equal to the amount of oxygen absorbed by the water.

Finally, the amount in milligrammes of oxygen absorbed thus found is multiplied by 4 for parts per million, and that result × 7 and ÷ 100 = grains per gallon.

10. Clark's Process for Hardness. Total ***before boiling*** **and permanent** ***after boiling.*** Solutions, etc., required.

(*a*) *Standard solution of Calcium chloride.*

Dissolve 8 decigrammes of pure crystallised calc-spar in dilute hydrochloric acid in a platinum dish, adding the acid gradually, and taking precautions to prevent loss by spurting. When all is dissolved, evaporate to dryness in a water bath, add a little distilled water and again evaporate to dryness. Repeat several times, to insure the expulsion of all the acid. Lastly, dissolve in water and make up to 700 c.c. For use, dilute to 10 times its volume: the result will be *standard water of 8° of hardness.* Or, instead of making the solution thus, dissolve ·1376 gramme pure crystallised selenite in 700 c.c. water, and that will be the *standard water of 8° of hardness.*

(*b*) *Standard Soap solution.*

Take 15 grammes lead plaster (*Emplastrum Plumbi,* B.P.), rub in a mortar with 4 grammes dry potassium carbonate; when fairly mixed add 660 c.c. of rectified methylated spirit, rub well, and after a time filter. To the filtrate add 440 c.c. of recently boiled distilled water, and mix. Now place 100 c.c. of the standard water (a) in an 8-ounce wide-mouthed stoppered bottle, and add from a burette the standard soap 1 c.c. at a time. After each addition, shake the bottle vigorously for about a quarter of a minute. As soon as a lather is produced, lay the bottle on its side after each addition, and observe if the lather remains permanent for five minutes. To ascertain this, at the end of five minutes roll the bottle half-way round; if the lather breaks instead of covering the whole surface of the water, it is not permanent; if it still covers the whole surface, it is permanent: now read the burette. Repeat the experiment, adding the full quantity of soap solution employed in the first experiment, less about 2 c.c.; shake as before, add soap solution very gradually till the permanent lather is formed, and note the number of c.c. of soap used. If this be exactly 18 c.c. then the soap is correct, but if less, then it must be diluted with proof spirit until 100 c.c. of standard water exactly takes 18 c.c. of soap. This method of dilution has been already discussed under Volumetric Analysis in Chapter VII. (see *Manufacture of Soda Solution,* page 112).

The Process.—(*a*) *For total hardness.* Put 100 c.c. of the water into the bottle and add the soap from a burette exactly as above described; note the number of c.c. of soap used, and refer to the following table:—

HARDNESS	C.C. OF SOAP USED	DIFFERENCE
0°	0·9	1·0
1°	2·9	2·5
2°	5·4	2·3
3°	7·7	2·2
4°	9·9	2·1
5°	12·0	2·0
6°	14·0	2.0
7°	16·0	2·0
8°	18·0	

and so on, after this point each 2 c.c. of soap used representing 1 additional degree of hardness.

Example to show use of this table.—Suppose 8 c.c. of soap to have been used, which comes between 7·7 and 9·9, then $8 - 7{\cdot}7 = {\cdot}7$; and $\frac{{\cdot}7}{2{\cdot}1} = {\cdot}33$; lastly ${\cdot}33 + 3° = 3{\cdot}33°$, correct degree of hardness.

If magnesium salts are present in the water the character of the lather will be very much modified, and a kind of scum (simulating a lather) will be seen in the water before the reaction is completed. The character of this scum must be carefully watched, and the soap test added more carefully, with an increased amount of shaking between each addition. With this precaution it will be comparatively easy to distinguish the point when the false lather due to the magnesium salt ceases, and the true persistent lather is produced. If the water is of more than 16° of hardness, mix 50 c.c. of the sample with an equal volume of recently boiled distilled water, which has been cooled in a closed vessel, and make the determination on this mixture of the sample and distilled water. In this case it will of course be necessary to multiply the figures obtained from the table by 2.

(*b*) *For permanent hardness.* To determine the hardness after boiling, boil a measured quantity of the water in a flask briskly for half an hour, adding distilled water from time to time to make up for loss by evaporation. It is not desirable to boil the water under a vertical condenser, as the dissolved carbonic acid is not so freely liberated. At the end of half an hour, allow the water to cool, the mouth of the flask being closed; make the water up to its original volume with recently boiled distilled water, and, if possible, decant the quantity necessary for testing. If this cannot be done quite clear, it must be filtered. Conduct the test in the same manner as described above.

The hardness is to be returned in each case to the nearest half-degree.

11. Judging the Results.

No definite rule can be laid down for judging all the results on one uniform scale, because the analyst must have special information as to the locality, nature of the soil, or depth of the well, before giving a reliable opinion. For example, nitrates, which have in river and shallow surface waters the highest significance, as indicating the presence of previous sewage contamination, entirely lose such force in waters from deep artesian wells, because these are naturally rich in such salts. The same thing may be said of ammonia, which, although highly unfavourable in shallow waters, is yet always found in artesian wells, most probably from the metal pipes acting as reducing agents upon the nitrates. Again, with upland peaty waters we always find a large reduction of permanganate, and consequently an excess of "oxygen consumed," although the organic matter so acting cannot be viewed as dangerous.

Setting aside, however, all questions of mineral constituents, and only looking at the indications of the presence or absence of organic matter, the author has devised a valuation scale, originally presented by him in a paper read before the Society, and which has proved since that time, in the hands of his various students dealing with sanitary examinations, as nearly correct as any general scale can be. The principle (which was originally proposed by the late Mr. Wigner) is to divide the amount of each figure found in the analysis by a fixed divisor, and *where the quotient exceeds* 10 *to double all figures over that number.* Let us suppose that, for example, a water yielded ·012 grain of albuminoid ammonia per gallon, and that the divisor fixed for this indication is ·0007; then we have

$$\frac{\cdot 012}{\cdot 0007} = 17{\cdot}1\text{; then } 17{\cdot}1 - 10 = 7{\cdot}1\text{, and } 7{\cdot}1 \times 2 = 14{\cdot}2\text{;}$$

therefore 14·2 + 10 = 24·2, indicated degree of impurity.

To prevent the production of enormous figures, likely to startle non-professional persons, the indicated degree of impurity is expressed as a decimal by dividing it by 100. Thus, it is only when the article is very bad indeed that the indication comes into full numbers.

Taking, then, the whole scale, it stands as follows :—

GRAINS PER GALLON.

Ammonia	each	·0015 = 1.
Albuminoid Ammonia	,,	·0007 – 1.
Oxygen consumed in 15 minutes	,,	·004 = 1.
Oxygen consumed in 4 hours	,,	·010 = 1.

PARTS PER MILLION.

Ammonia	each	·02 = 1.
Albuminoid Ammonia	,,	·01 = 1.
Oxygen in 15 minutes	,,	·057 –
Oxygen in 4 hours	,,	·143 = .

When any number exceeds 10, then all over 10 is to be doubled and added to the original number, and the total valuation is to be divided by 100 and noted as "comparative degree of organic impurity." Then, *supposing no other consideration intervenes to modify the analyst's opinion of the sample*, the following limits should be observed :—

1st Class Water	*up to* ·25 degree.
2nd ,, ,, (more or less questionable)	*up to* ·40 ,,
Undrinkable Water	*over* ·40 ,,

DIVISION II. THE SANITARY ANALYSIS OF AIR.

For sanitary purposes it is not necessary to make a full analysis of air, but should such be desired the processes described in Chapter XII. must be followed. The chief points are :—

1. Estimation of Carbon Dioxide.

This is done by the method of Pettenkofer, which consists in standardising 100 c.c. of lime or baryta water with standard oxalic acid 2·25 grammes per litre of which 1 c.c. = ·001 (1 milligramme of CaO). The air to be examined having been collected in a large bottle of known capacity, 100 c.c. of the same lime-water are added, the bottle is closed and well shaken for some time. The CO_2 is absorbed, forming $CaCO_3$. The resulting milky liquid is titrated in the bottle with the same acid. The indicator used is turmeric paper, or phenol-phthalein. The difference between the two titrations gives the amount of CaO precipitated as carbonate by the CO_2 in the air, and this is then calculated thus :—

$$\frac{\text{c.c. used} \times \cdot 001 \times 44}{56} = CO_2 \text{ present in the volume of air taken.}$$

In strict analyses, the volume of air taken must be corrected for observed temperature and pressure to its volume at N.T.P. Pure air contains about ·04 per cent. of CO_2.

2. Estimation of Organic Matter.

A known volume of air is sucked by an aspirator through a specially arranged apparatus containing ammonia-free distilled water, and the resulting liquid is analysed for "free" and "albuminoid" ammonia like a water.

3. Testing for Gaseous Impurities.

The odor will generally detect these when present in notable proportions. Blotting-paper dipped (*a*) in tincture of turmeric, turns red-brown in presence of ammonia; (*b*) in solution of subacetate of lead—black with sulphuretted hydrogen; (*c*) in solution potassium iodide mixed with starch paste—blue with chlorine or ozone; (*d*) weak solution of indigo is decolorised by chlorine and sulphurous acid gas; (*e*) red litmus paper dipped in solution of potassium iodide becomes blue with ozone and not with chlorine.

DIVISION III. FOOD ANALYSIS.

Here we will only attempt to consider a few of the more commonly occurring cases, always choosing the simplest and most rapid process.

1. Milk.

(1) **Specific Gravity.** Take the specific gravity with great care at 60° F. If not at 60°, take the temperature and refer to the table opposite.

(2) **Total Solids.** Heat a small flat platinum dish about 1½ inch in diameter to redness, cool it under the desiccator and weigh. Put in about 5 c.c. of the milk and again weigh. The difference = milk taken. Now dry on the water bath for 3 hours (or until seemingly dry), then transfer to the drying oven at 212° for 2 hours, cool under the desiccator and weigh. Put it back in the oven for an hour, repeat the cooling and weighing, and if the difference does not exceed a milligramme or two it is dry, if it does, then repeat the drying. The weight of the dry residue *minus* the tare of the dish equals the total solids; then—

$$\frac{\text{weight solids} \times 100}{\text{quantity taken}} = \text{per cent. of total solids.}$$

(3) **Fat.** May now be calculated by Fleishmann's formula thus (t = total solids: s = specific gravity: f = fat):—

$$f = t \cdot 833 - 2 \cdot 2 \frac{100s - 100}{s}$$

Deducting the fat thus found from the total solids, we get the "solids not fat."

If a more accurate process be required, the fat may be estimated by Adams' method as modified by Thompson thus:—

A strip of good filtering paper (not blotting) is procured, 21 in. long by 2½ in. wide. Two small pieces of ordinary stirring rod are taken, the one rather longer than the other. These are fixed together by stretching a section of indiarubber tubing over each end of the rods. The two rods are separated from each other by the fingers, and the end of the strip of filter paper placed between them. The longer rod may be conveniently held by an iron clamp on a retort stand. 5 c.c. of the milk are taken in a pipette, having a long stem under the bulb, and whilst the left hand holds the free end of the strip of paper, thus giving it an almost horizontal position, a small portion of the milk is allowed to run from the pipette which is held in the right hand, with the finger closing the top, so as to make a line of milk across the strip of paper within about an inch of the fixed end. More milk is then allowed to flow on to the centre of the strip, and is then spread equally all over the surface with the stem of the pipette, which is held almost horizontally at right angles to the strip. By this means the milk may be transferred entirely to the strip of paper, about an inch or an inch and a half at the end being left unmoistened, upon which the stem of the pipette is wiped dry. The strip of paper thus moistened with 5 c.c. of milk may now be taken between the hands, and dried in 2 or 3 minutes over an ordinary Bunsen burner. The flame may be allowed to play directly on the paper, which is moved rapidly backwards and forwards over it. The strip of paper thus treated is then coiled on a stirring rod, the rod withdrawn, and the coil without any further manipulation placed in a Soxhlet's tube for extracting the fat (see Chapter I., page 2).

Observed Specific Gravity.	DEGREES OF THERMOMETER (*Fahr.*).																					Observed Specific Gravity.
	50	51	52	53	54	55	56	57	58	59	60	61	62	63	64	65	66	67	68	69	70	
1·020·0	19·2	19·3	19·4	19·4	19·5	19·6	19·7	19·8	19·9	19.9	1·020·0	20·1	20·2	20·2	20·3	20·4	20·5	20·6	20·7	20·9	21·0	1·020·0
1·021·0	20·2	20·3	20·3	20·4	20·5	20·6	20·7	20·8	20·9	20·9	1·021·0	21·1	21·2	21·3	21·4	21·5	21·6	21·7	21·8	22·0	22·1	1·021·0
1·022·0	21·2	21·3	21·3	21·4	21·5	21·6	21·7	21·8	21·9	21·9	1·022·0	22·1	22·2	22·3	22·4	22·5	22·6	22·7	22·8	23·0	23·1	1·022·0
1·023·0	22·2	22·3	22·3	22·4	22·5	22·6	22·7	22·8	22·8	22·9	1·023·0	23·1	23·2	23·3	23·4	23·5	23·6	23·7	23·8	24·0	24·1	1·023·0
1·024·0	23·2	23·3	23·3	23·4	23·5	23·6	23·6	23·7	23·8	23·9	1·024·0	24·1	24·2	24·3	24·4	24·5	24·6	24·7	24·9	25·0	25·1	1·024·0
1·025·0	24·1	24·2	24·3	24·4	24·5	24·6	24·6	24·7	24·8	24·9	1·025·0	25·1	25·2	25·3	25·4	25·5	25·6	25·7	25·9	26·0	26·1	1·025·0
1·026·0	25·1	25·2	25·2	25·3	25·4	25·5	25·6	25·7	25·8	25·9	1·026·0	26·1	26·2	26·3	26·5	26·6	26·7	26·8	27·0	27·1	27·2	1·026·0
1·027·0	26·1	26·2	26·2	26·3	26·4	26·5	26·6	26·7	26·8	26·9	1·027·0	27·1	27·3	27·4	27·5	27·6	27·7	27·8	28·0	28·1	28·2	1·027·0
1·028·0	27·0	27·1	27·2	27·3	27·4	27·5	27·6	27·7	27·8	27·9	1·028·0	28·1	28·3	28·4	28·5	28·6	28·7	28·8	29·0	29·1	29·2	1·028·0
1·029.0	28·0	28·1	28·2	28·3	28·4	28·5	28·6	28·7	28·8	28·9	1·029·0	29·1	29·3	29·4	29·5	29·6	29·8	29·9	30·1	30·2	30·3	1·029·0
1·030·0	29·0	29·1	29·1	29·2	29·3	29·4	29·6	29·7	29·8	29·9	1·030·0	30·1	30·3	30·4	30·5	30·7	30·8	30·9	31·1	31·2	31·3	1·030·0
1·031·0	29·9	30·0	30·1	30·2	30·3	30·4	30·5	30·6	30·8	30·9	1·031·0	31·2	31·3	31·4	31·5	31·7	31·8	32·0	32·2	32·2	32·4	1·031·0
1·032·0	30·9	31·0	31·1	31·2	31·3	31·4	31·5	31·6	31·7	31·9	1·032·0	32·2	32·3	32·5	32·6	32·7	32·9	33·0	33·2	33·3	33·4	1·032·0
1·033·0	31·8	31·9	32·0	32·1	32·3	32·4	32·5	32·6	32·7	32·9	1·033·0	33·2	33·3	33·5	33·6	33·8	33·9	34·0	34·2	34·3	34·5	1·033·0
1·034·0	32·7	32·9	33·0	33·1	33·2	33·3	33·5	33·6	33·7	33·9	1·034·0	34·2	34·3	34·5	34·6	34·8	34·9	35·0	35·2	35·3	35·5	1·034·0
1·035·0	33·6	33·8	33·9	34·0	34·2	34·3	34·5	34·6	34·7	34·9	1·035·0	35·2	35·3	35·5	35·6	35·8	35·9	36·1	36·2	36·4	36·5	1·035·0

(4) **Added Water.** The limit for the strength of milk is at present based upon that of the poorest possible natural milk. Average milk will show:—

Fat	. .	3·00
Solids not fat	. .	9·00
	Total	12·00.

If, however, a milk has only—

Fat	.	3·0
Solids not fat	.	8·5
	Total	11·5,

it will not be considered as definitely proved to be adulterated. The amount of water added should, however, always be calculated upon the average standard of 9 per cent. "solids not fat," provided the milk is certainly well below the limit of 8·5%. The "solids not fat" are used for the basis of calculation because they are a fairly constant quantity, the fat being variable. The amount of pure standard milk present in any sample may be calculated thus:—

$$\frac{\text{solids not fat} \times 100}{9} = \%\ \text{of pure milk present,}$$

and the difference between this result and 100 is of course added water.

(5) **Ash.** The total solids in the platinum dish are burned over a low flame at dull redness till quite white, and the ash is weighed. The ash should be about ·70%, and it will never, as a rule, fall below ·67 in an unwatered milk.

2. Butter.

Undoubtedly the best process for the detection of foreign fat in butter is:—

Reichert's Process.—This process is based upon the presence in butter-fat of one constituent (tributyrin), which yields by appropriate treatment an acid (butyric acid) that is relatively much more volatile than the other acids yielded by any of the practicable substitutes for butter-fat. As will be seen, the method is based on operations that admit of no arbitrary variation; to secure reliable and comparable results exactly similar steps must be followed by all operators.

2·5 grammes of the filtered melted fat are weighed into a flask of about 150 c.c. capacity, 20 c.c. of a solution of potassium hydrate in methylated spirit—50 grammes KHO in 1000 c.c.—and the whole heated to gentle ebullition on a water bath until the fat is entirely saponified and all the alcohol expelled. The soap should form an almost dry mass, that can scarcely be detached from the bottom of the flask by shaking; the last traces of alcohol being removed by occasionally sucking the air out of the flask with a tube. Afterwards 50 c.c. of water is added to dissolve the soap (the solution is hastened by *gentle* heating), and when the soap is completely dissolved, 20 c.c. of dilute sulphuric acid (100 c.c. H_2SO_4 in 1000 c.c. of water) is added to decompose the soap. The flask is then connected with a small but efficient Liebig's condenser, and the contents heated to moderate boiling, with addition of two or three bits of broken tobacco pipe to prevent bumping. The distillate, which contains some insoluble acids, must be passed, as it drops from the condenser, through a small wet filter into a 50 c.c. measure. The distillation is continued until exactly 50 c.c. has come over, which is at once titrated with decinormal soda solution, phenol-phthalein being used as the indicator.

Reichert's formula for determining the percentage of butter-fat in mixed fat is $B = 7·3\ (n - 0·3)$; n being the number of cubic centimetres of decinormal alkali used in neutralising the distillate from 2·5 grammes of the fat.

Thus treated, pure butter never yields less acidity than is represented by

12 c.c. of decinormal soda (4 grammes *real* NaHO per litre). Butter made from the milk of one single individual cow has been known to fall to 11·5 c.c., but average butter, as produced from the mixed milk of a herd of cows, takes 14 c.c. Every time a fresh lot of alcoholic potash (1 in 20) is made, a blank experiment must be gone through without any fat, and the amount of alkali used must be noted on the potash bottle and deducted as a check from each actual analysis. This is necessary when we employ methylated spirit, and such check may range up to 2 c.c. with certain spirit.

3. Bread.

The only common impurity in bread detectable by chemical means is alum, which is tested for as follows: A piece of the crumb of bread, cut from the centre of the loaf, is steeped in a mixture of 5 c.c. freshly made tincture of logwood, 5 c.c. of solution of ammonium carbonate, and 20 c.c. of water. The liquid is poured off and the basin placed on the top of the water oven. If alum be present a fine blue color will be developed. This test is, however, only very strong presumptive evidence; and to absolutely confirm it in legal cases 100 grammes of the bread should be dried and incinerated in a large platinum basin, the ash dissolved in hydrochloric acid, and the amount of alumina present estimated by the ordinary methods of quantitative analysis. Proper precautions should be taken for separating silica and iron, and an allowance should be made on the alumina found from the amount of silica present, to avoid the estimation of Al_2O_3 accidentally present as clay. The alumina is best precipitated and weighed as phosphate in the presence of ammonium acetate and excess of acetic acid, the solution being boiled and filtered hot.

4. Estimation of the Alcoholic strength of Spirits, Beer, Wines, and Tinctures.

(1) **In a Pure Spirit.** If the sample be simply a dilute spirit which leaves no residue upon evaporation, the percentage may be ascertained by taking the specific gravity. Great care must be taken that the contents of the specific gravity bottle are exactly at a temperature of 60° F., and in taking such specific gravities it is better to perform the operation, say, three times, and take the average of such determinations, as a very small error makes a great difference in the commercial value of the sample under examination. Reference to the alcohol table appended will now give the required information.

(2) **In a Wine, Beer, Tincture, or Colored or Sweetened Spirit.** The specific gravity of the sample is taken at 60° F. and noted. 100 c.c. are measured off at 60° F., and evaporated on the water bath or over a low gas flame, so as just to boil very gently, till all odor of alcohol has passed off. The liquid thus left is poured back into the measuring flask, the beaker is rinsed with a little distilled water, and the rinsings added to the flask. The whole is cooled to 60° F., made up to the 100 c.c. mark with distilled water, and the specific gravity of this non-alcoholic fluid is then taken, also at 60° F. Lastly, we calculate :—

$$\frac{\text{Gravity before boiling}}{\text{Gravity after boiling}} = \text{true specific gravity of the contained spirit.}$$

The gravity so found gives by the table the percentage of alcohol in the sample.

(3) Table for ascertaining the percentages respectively of Alcohol by Weight, by Volume, and as Proof Spirit, from the Specific Gravity.

Condensed from the excellent Alcohol Tables of Mr. Hehner in the "Analyst," vol. v., pp. 43—63.

Specific gravity, 15·5°.	Absolute Alcohol by w'ght. Per cent.	Absolute Alcohol by vol'me Per cent.	Proof Spirit. Per cent.
1·0000	0·00	0·00	0·00
·9999	0·05	0·07	0·12
·9989	0·58	0·73	1·28
·9979	1·12	1·42	2·48
·9969	1·75	2·20	3·85
·9959	2·33	2·93	5·13
·9949	2·89	3·62	6·34
·9939	3·47	4·34	7·61
·9929	4·06	5·08	8·90
·9919	4·69	5·86	10·26
·9909	5·31	6·63	11·62
·9899	5·94	7·40	12·97
·9889	6·64	8·27	14·50
·9879	7·33	9·13	15·99
·9869	8·00	9·95	17·43
·9859	8·71	10·82	18·96
·9849	9·43	11·70	20·50
·9839	10·15	12·58	22·06
·9829	10·92	13·52	23·70
·9819	11·69	14·46	25·34
·9809	12·46	15·40	26·99
·9799	13·23	16·33	28·62
·9789	14·00	17·26	30·26
·9779	14·91	18·36	32·19
·9769	15·75	19·39	33·96
·9759	16·54	20·33	35·63
·9749	17·33	21·29	37·30
·9739	18·15	22·27	39·03
·9729	18·92	23·19	40·64
·9719	19·75	24·18	42·38
·9709	20·58	25·17	44·12
·9699	21·38	26·13	45·79
·9689	22·15	27·04	47·39
·9679	22·92	27·95	48·98
·9669	23·69	28·86	50·57
·9659	24·46	29·76	52·16
·9649	25·21	30·65	53·71
·9639	25·93	31·48	55·18
·9629	26·60	32·27	56·55
·9619	27·29	33·06	57·94
·9609	28·00	33·89	59·40
·9599	28·62	34·61	60·66
·9589	29·27	35·35	61·95
·9579	29·93	36·12	63·30
·9569	30·50	36·76	64·43
·9559	31·06	37·41	65·55
·9549	31·69	38·11	66·80
·9539	32·31	38·82	68·04
·9529	32·94	39·54	69·29
·9519	33·53	40·20	70·46
·9509	34·10	40·84	71·58
·9499	34·57	41·37	72·50
·9489	35·05	41·90	73·43
·9479	35·55	42·45	74·39
·9469	36·06	43·01	75·37
·9459	36·61	43·63	76·45
·9449	37·17	44·24	77·53
·9439	37·72	44·86	78·61
·9429	38·28	45·47	79·68
·9419	38·83	46·08	80·75
·9409	39·35	46·64	81·74
·9399	39·85	47·18	82·69
·9389	40·35	47·72	83·64
·9379	40·85	48·26	84·58
·9369	41·35	48·80	85·53
·9359	41·85	49·34	86·47
·9349	42·33	49·86	87·37
·9339	42·81	50·37	88·26
·9329	43·29	50·87	89·15
·9319	43·76	51·38	90·03
·9309	44·23	51·87	90·89
·9299	44·68	52·34	91·73
·9289	45·14	52·82	92·56
·9279	45·59	53·29	93·39
·9269	46·05	53·77	94·22
·9259	46·50	54·24	95·05
·9249	46·96	54·71	95·88
·9239	47·41	55·18	96·70
·9229	47·86	55·65	97·52
·9219	48·32	56·11	98·34
·9209	48·77	56·58	99·16
·9199	49·20	57·02	99·93
·9198	49·24	57·06	100·00Ps
·9189	49·68	57·49	100·76
·9179	50·13	57·97	101·59
·9169	50·57	58·41	102·35
·9159	51·00	58·85	103·12
·9149	51·42	59·26	103·85
·9139	51·83	59·68	104·58
·9129	52·27	60·12	105·35
·9119	52·73	60·56	106·15
·9109	53·17	61·02	106·93
·9099	53·61	61·45	107·69
·9089	54·05	61·88	108·45
·9079	54·52	62·36	109·28
·9069	55·00	62·84	110·12
·9059	55·45	63·28	110·92
·9049	55·91	63·73	111·71
·9039	56·36	64·18	112·49
·9029	56·82	64·63	113·26
·9019	57·25	65·05	113·99
·9009	57·67	65·45	114·69
·8999	58·09	65·85	115·41
·8989	58·55	66·29	116·18
·8979	59·00	66·74	116·96
·8969	59·43	67·15	117·68
·8959	59·87	67·57	118·41
·8949	60·29	67·97	119·12
·8939	60·71	68·36	119·80
·8929	61·13	68·76	120·49
·8919	61·54	69·15	121·18
·8909	61·96	69·54	121·86
·8899	62·41	69·96	122·61
·8889	62·86	70·40	123·36
·8879	63·30	70·81	124·09
·8869	63·74	71·22	124·80
·8859	64·17	71·62	125·51
·8849	64·61	72·02	126·22
·8839	65·04	72·42	126·92
·8829	65·46	72·80	127·59
·8819	65·88	73·19	128·25
·8809	66·30	73·57	128·94
·8799	66·74	73·97	129·64
·8789	67·17	74·37	130·33
·8779	67·58	74·74	130·98
·8769	68·00	75·12	131·64
·8759	68·42	75·49	132·30
·8749	68·83	75·87	132·95
·8739	69·25	76·24	133·60
·8729	69·67	76·61	134·25
·8719	70·08	76·98	134·90
·8709	70·48	77·32	135·51
·8699	70·88	77·67	136·13
·8689	71·29	78·04	136·76
·8679	71·71	78·40	137·40
·8669	72·13	78·77	138·05
·8659	72·57	79·16	138·72
·8649	73·00	79·54	139·39
·8639	73·42	79·90	140·02
·8629	73·83	80·26	140·65
·8619	74·27	80·64	141·33
·8609	74·73	81·04	142·03
·8599	75·18	81·44	142·73
·8589	75·64	81·84	143·42
·8579	76·08	82·23	144·10
·8569	76·50	82·58	144·72
·8559	76·92	82·93	145·34
·8549	77·33	83·28	145·96
·8539	77·75	83·64	146·57
·8529	78·16	83·98	147·17
·8519	78·56	84·31	147·75
·8509	78·96	84·64	148·32
·8499	79·36	84·97	148·90
·8489	79·76	85·29	149·44
·8479	80·17	85·63	150·06
·8469	80·58	85·97	150·67
·8459	81·00	86·32	151·27
·8449	81·40	86·64	151·83
·8439	81·80	86·96	152·40
·8429	82·19	87·27	152·95
·8419	82·58	87·58	153·48
·8409	82·96	87·88	154·01
·8399	83·35	88·19	154·54
·8389	83·73	88·49	155·07
·8379	84·12	88·79	155·61
·8369	84·52	89·11	156·16
·8359	84·92	89·42	156·71
·8349	85·31	89·72	157·24
·8339	85·69	90·02	157·76
·8329	86·08	90·32	158·28
·8319	86·46	90·61	158·79
·8309	86·85	90·90	159·31
·8299	87·23	91·20	159·82
·8289	87·62	91·49	160·33
·8279	88·00	91·78	160·84
·8269	88·40	92·08	161·37
·8259	88·80	92·39	161·91
·8249	89·19	92·68	162·43
·8239	89·58	92·97	162·93
·8229	89·96	93·26	163·43
·8219	90·32	93·52	163·88
·8209	90·68	93·77	164·33
·8199	91·04	94·03	164·78
·8189	91·39	94·28	165·23
·8179	91·75	94·53	165·67
·8169	92·11	94·79	166·12
·8159	92·48	95·06	166·58
·8149	92·85	95·32	167·04
·8139	93·22	95·58	167·50
·8129	93·59	95·84	167·96
·8119	93·96	96·11	168·24
·8109	94·31	96·34	168·84
·8099	94·66	96·57	169·24
·8089	95·00	96·80	169·65
·8079	95·36	97·05	170·07
·8069	95·71	97·29	170·50
·8059	96·07	97·53	170·99
·8049	96·40	97·75	171·30
·8039	96·73	97·96	171·68
·8029	97·07	98·18	172·05
·8019	97·40	98·39	172·43
·8009	97·73	98·61	172·80
·7999	98·06	98·82	173·17
·7989	98·37	99·00	173·50
·7979	98·69	99·18	173·84
·7969	99·00	99·37	174·17
·7959	99·32	99·57	174·52
·7949	99·65	99·77	174·87
·7939	99·97	99·98	175·22
Absolute Alcohol.			
·7938	100·00	100·00	175·25

5. Mustard.

This is chiefly a microscopical matter for the exact identification of impurities, but the following chemical operations may be performed :—

(1) Test a cooled decoction for starch with solution of iodine.

(2) If starch be found, extract a weighed portion in the "Soxhlet" with petroleum spirit or ether. Distil off the spirit dry, and weigh the oil. Mustard contains as an ordinary minimum 33 % of oil, and the amount of genuine mustard in the sample will then be found thus :—

$$\frac{\%\text{ of oil found} \times 100}{33} = \%\text{ genuine mustard};$$

by deducting this from 100 the difference is added starch or flour.

(3) Moisten the mustard with a little ammonia, when the turmeric brown will be developed if that coloring agent be present.

6. Pepper.

This is also examined microscopically for extraneous starches (especially rice), and also for *poivrette*, which is chiefly ground olive stones.

(1) Dry at 212°, weigh out 5 grammes of the dried pepper, and take the ash. This should not exceed 9·0 per cent. even in the most inferior black pepper, which has been *previously dried at* 212°. Treat the ash with HCl, add water, boil, filter, wash, dry, ignite, and weigh the sand. This should not exceed 4 per cent. on the dried pepper.

(2) If the microscope reveals starch other than that of pepper, estimate it on 2 grammes of the dried pepper by the method given in Section 8.

7. Colored Sweets.

The poisonous colors are nearly all mineral and insoluble. They may be scraped off, washed with water, and identified by the ordinary methods given in Chapter IV. As a rule, at present only aniline colors are used, and they are added in such minute proportions as not to be considered dangerous.

8. Direct Estimation of Starch in Cereals, Mustard, Pepper, and all Foods and Drugs containing it.

This depends upon the fact that starch forms an insoluble compound with barium. If an excess of baryta water of known strength be added to starch which has been gelatinised in water, a portion of the barium will combine with the starch, and then by estimating the excess of baryta water left unabsorbed we can find the amount taken up by the starch. The formula of the starch-baryta compound is $C_{24}H_{40}O_{20}$ BaO, and it therefore contains 19·1 per cent. of BaO.

The materials required are :—

(1) Decinormal hydrochloric acid, containing 3·65 grms. real HCl per 1000 c.c. = ·00765 BaO for each c.c.

(2) Baryta water kept in a special jar with a burette permanently attached, as shown in the illustration (overleaf). A is the jar for the baryta water, having a tube attached containing lumps of quicklime to prevent the entrance of CO_2 from the air. The burette B is attached to the bottom neck of the jar by a tube having a pinchcock (*a*) to admit the reagent, and a tube (*n*) filled with lumps of caustic potash to prevent entrance of CO_2.

(3) Alcoholic solution of phenol-phthalein as an indicator.

The Process.—The substance is powdered or finely ground in a mill, and 3 grammes weighed out for analysis. If (as in the case of mustard, pepper, etc.) it contains oil or resinous bodies, these are first extracted by percolation with petroleum spirit or ether in the "Soxhlet." The powder is then well rubbed in with successive quantities of water until thoroughly disintegrated, the liquid being transferred to a 250 c.c. flask, and 100 c.c. of water in all being used to entirely transfer the powder from the mortar to the flask. In dealing with very hard substances, like beans, peas, etc., the water should be used boiling. The flask and contents are now heated on the water bath for half an hour, with frequent shaking, to *entirely gelatinise* the starch. The whole is then cooled, and 50 c.c. of standard baryta water having been added from the burette, the flask is corked, well shaken for two minutes, proof spirit is added up to the 250 c.c. mark, and the whole again shaken, *tightly corked*, and set

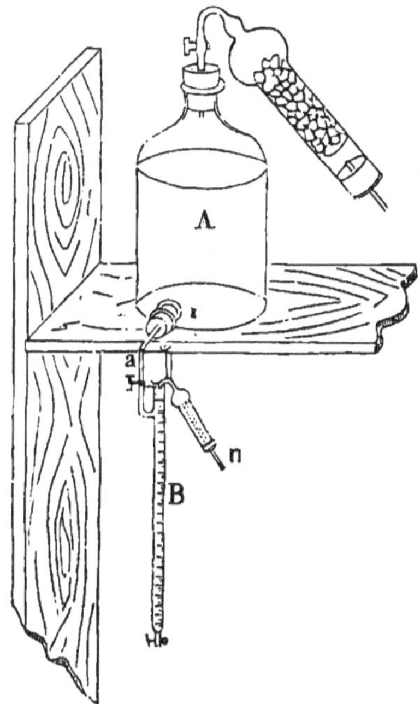

Fig. 45.

aside to settle. While settling a check is made on 50 c.c. of the baryta water by shaking up with 100 c.c. of freshly boiled distilled water in a 250 c.c. corked flask, and then titrating with the decinormal HCl, in the presence of two drops of phenol-phthalein. The number of c.c. of acid used is recorded, giving the total strength of 50 c.c. of the baryta water employed. When the main analysis has settled, 50 c.c. of the clear liquid are drawn off with a pipette, rapidly titrated with the decinormal acid and phenol-phthalein, as in the check, and the number of c.c. of acid used is multiplied by 5 and set down as strength of baryta water remaining uncombined. This latter amount is deducted from the total check strength, and the difference in c.c. of acid is multiplied by ·00765, which gives BaO combined with the starch. This result now multiplied by 4·2353 gives the amount of starch in the 3 grms. taken $= x$: then

$$\frac{x \times 100}{3} = \text{percentage of starch.}$$

9. Free Sulphuric Acid in Vinegar.

(1) *B. P. Process.*—If ten minims of solution of barium chloride be added to a fluid ounce of the vinegar, and the precipitate, if any, be separated by filtration, a further addition of the test will give no precipitate (absence of more than $\frac{1}{10}$ per cent. of sulphuric acid).

(2) *Improved Method.*—The official test given above is entirely valueless in the case of vinegars containing much sulphate, from being made with water rich in that ingredient; it is therefore necessary, if the vinegar appear to be adulterated when thus tried, to proceed as follows: 50 c.c. of the vinegar are mixed with 25 c.c. of volumetric solution of sodium hydrate, made decinormal by diluting the official volumetric solution to ten times its bulk with water. The whole is evaporated to dryness, and incinerated at the lowest possible temperature. 25 c.c. of decinormal solution of oxalic acid (made to exactly balance the sodium hydrate solution) are now added to the ash, the liquid heated to expel CO_2, and filtered. The filter is washed with hot water, and the washings having been added to the filtrate, litmus solution is added, and the amount of free acid ascertained by running in decinormal soda from a burette. The number of c.c. of soda thus used multiplied by ·0049 gives the amount of free sulphuric acid in the vinegar. This process depends on the fact that *whenever the ash of vinegar has not an alkaline reaction, free mineral acid was undoubtedly added.*

CHAPTER XI.

ANALYSIS OF DRUGS, URINE AND URINARY CALCULI.

DIVISION I. ANALYSIS OF DRUGS.

1. General Scheme.

THE analysis of drugs is so large a subject that only a few of the more commonly occurring problems can be discussed in the present volume. It will, however, be interesting, before proceeding to the consideration of special matters, to give a sketch of the general method of analysing a vegetable substance used in medicine, following the lines laid down by Dragendorff, but somewhat modified by the author as the result of experience.

Step I. Dry the substance in the water oven until it ceases to lose weight.

Step II. Pack the dried and powdered substance, mixed with a little sand, in a "Soxhlet" apparatus, thoroughly exhaust it with petroleum spirit, and cork up and save the fluid extract so obtained, marking it A.

Step III. Spread the solid left from Step II. out to dry on a plate of glass on the top of the water oven, and when all odor of petroleum has passed off, replace it in the Soxhlet, and exhaust it this time with perfectly anhydrous ether. Cork up the ethereal extract obtained and mark it B.

Step IV. Spread out as before, and when all odor of ether is gone, repack and extract with purified commercial methyl alcohol, as sold for making methylated spirit. This is more volatile than common alcohol, and is as a rule a better solvent of the articles required in this group, while it does not so readily extract glucose, etc. It is an article of commerce, and can be specially ordered through a purveyor of chemicals, as "commercial methol, highest strength."—Save this alcoholic extract and mark it C.

Step V. Extract the insoluble matter from Step IV. with distilled water at a temperature not exceeding 120° Fah., and filter. Wash with cold water and save the filtrate D.

Step VI. Wash the insoluble off the filter into a large flask, with plenty of water, acidulated with 1 per cent. of hydrochloric acid, and boil it for an hour under an upright condenser. Let it settle, pour off the liquid as close as possible (saving it), and then collect the insoluble on a filter and wash with boiling water, adding the washings to what was poured off. This extract is marked E.

Step VII. Once more wash the insoluble from the filter into a beaker and boil it up for an hour with plenty of water rendered distinctly alkaline with sodium hydrate. Collect on a weighed filter, wash first with boiling water, acidulated with hydrochloric acid, and then with plain boiling water, till no trace of a chloride remains; dry in the water oven and weigh, deducting the tare of the filter. Lastly, ignite the filter and its contents in a weighed platinum basin, and deduct the ash so found from the first weight, and the difference will be the woody fibre in the drug.

Step VIII. Make a *nitrogen* determination by Kjeldahl's method (page 154) on a fresh portion, and the nitrogen found (after deducting any due to alkaloids present) multiplied by 6·33 will give the amount of albuminous bodies present.

Treatment of the Separate Solutions.—Each liquid is made to a definite number of c.c. with the same solvent, and then an aliquot part, say 10 c.c., is taken and evaporated, and the residue weighed, to find the total matter soluble in each solvent. The bulk of the liquids are then treated as follows:—

Liquid A. This will contain chiefly fixed and volatile oils. The spirit is allowed to evaporate spontaneously, and the residue is distilled with water, when the volatile oil passes over, leaving the fixed oil in the retort.

Liquid B. This chiefly contains resins, together with some bitters, alkaloids and organic acids. The solution is evaporated to dryness on the water.bath with sand, and the residue, having been powdered, is boiled with water slightly acdulated with HCl. A portion of this watery solution is tested for benzoic, cinnamic, salicylic, gallic, and other free organic acids, and the remainder is saved for subsequent use in Group C. The portion insoluble in water now chiefly represents any resins present in the drug, which are soluble in ether. These may be further divided and examined by the action of alcohol. Resins are recognised by their odor on warming, and by the action of H_2SO_4, HNO_3, HCl, etc., on spots of the solid resin left by evaporating the solutions. This matter requires special experience, but a full description of the nature and reaction of all the principal resins will be found in the author's work on "Pharmaceutical and Medical Chemistry," or in his book on "Organic Materia Medica."

Liquid C. Is evaporated to a low bulk, and then poured into water faintly acidulated with hydrochloric acid. Any insoluble matter is probably a resinous body insoluble in ether, and is to be filtered out and examined as a resin. A portion of the aqueous solution is to be tested for tannin, and the remainder is to be mixed with the reserved liquid from B, the whole gently evaporated to a convenient bulk and treated by immiscible solvents as follows:—

Step I. The liquid (which must still retain a slightly acid reaction) is shaken up successively with chloroform and ether in a separator (see fig. 23, page 93). The solvents are drawn off and evaporated, and the residues so obtained tested for glucosides and bitter principles, of which latter bodies the following are some of the more commonly occurring :—

(1) Extracted by chloroform from acid solutions :—

Absinthin (wormwood).
Anthemin (chamomiles).
Colchicine (colchicum), imperfectly.
Colocynthin (colocynth, or bitter apple), imperfectly.
Calumbin, and probably some berberine (calumba), bright yellow, and highly fluorescent.
Gentipicrin (gentian), very imperfectly.
Picric acid (artificial), yellow, imperfectly.
Picrotoxin (*cocculus indicus*), with difficulty.
Quassiin (quassia wood).

(2) Subsequently extracted by ether from acid solutions :—

Chiratin (chiretta).
Colocynthin (colocynth, or bitter apple).
Gentipicrin (gentian).
Picric acid. yellow.
Picrotoxin (*cocculus indicus*).

Note.—The alkaloid colchicine comes out with the glucoside in this division.

Step II. The liquid remaining in the separator is now rendered alkaline with sodium hydrate and shaken up again with chloroform. This extracts nearly all the alkaloids. The chloroform is evaporated and the residue tested for alkaloids (see Chap. V.).

Step III. The liquid still remaining in the separator is shaken up with amylic alcohol, which takes out morphine and leaves it on evaporation, when any residue is tested for its presence.

Liquid D. Is evaporated to a low bulk and then mixed with twice its volume of rectified spirit, when gums precipitate insoluble, and may be examined, and sugars dissolve and may be estimated by "Fehling." Saponin also may be found with the sugars.

2. Analysis of Cinchona Bark.

Step I. Extraction. Mix 20 grammes (or 200 grains) of the bark in fine powder (No. 60) with 6 grammes (or 60 grains) of calcium hydrate in a mortar. Slightly moisten with 20 c.c. (or half a fluid ounce) of water, and mix intimately; allow the mixture to stand for an hour or two, when it will present the characters of a moist, dark brown powder, in which there should be no lumps or visible white particles. Transfer this powder to a 6-ounce flask, add 130 c.c. (or 3 fluid ounces) of benzolated amylic alcohol, boil them together for about half an hour under an upright condenser, decant and drain off the liquid on to a filter, leaving the powder in the flask; add more of the benzolated amylic alcohol to the powder, boil and decant as before; repeat this operation a third time; then turn the contents of the flask on to the filter, and wash by percolation with more of the benzolated amylic alcohol until the bark is exhausted. (If an upright condenser be not available, a funnel may be placed in the mouth of the flask, and another flask filled with cold water placed in the funnel. This will form a convenient condenser, which will prevent the loss of more than a small quantity of the boiling liquid.) Introduce the collected filtrate, while still warm, into a stoppered glass separator; add to it 2 c.c. (or 20 minims) of diluted hydrochloric acid, mixed with 12 c.c. (or 2 fluid drachms) of water; shake them well together, and when the acid liquid has separated draw it off, and repeat the process with distilled water slightly acidulated with hydrochloric acid, until the whole of the alkaloids have been removed (which is known by a few drops of the liquid ceasing to give any precipitate with sodium hydrate). The acid liquid thus obtained will contain all the alkaloids as hydrochlorates, and excess of hydrochloric acid.

Step II. Separation of the Quinine and Cinchonidine from the Quinidine, Cinchonine, and Amorphous Alkaloids.—The acid fluid from Step I. is to be carefully and exactly neutralised with ammonia while warm, and then concentrated to the bulk of 18 c.c. (or 3 fluid drachms). 1·5 gramme (or 15 grains) of tartrated soda, dissolved in twice its weight of water, is added to the neutral liquid, and the mixture stirred with a glass rod. Insoluble tartrates of quinine and cinchonidine will separate completely in about an hour; and these collected on a filter, washed, and dried in the water oven, will contain eight-tenths of their weight of the alkaloids, quinine and cinchonidine, which multiplied by 5 (or if grains have been taken, divided by 2) represents the percentage of those alkaloids. The other alkaloids will be left in the mother-liquor.

Step III. For total Alkaloids.—To the mother-liquor from the preceding process add solution of ammonia in slight excess. Collect, wash, and dry the precipitate in the water oven, which will contain the other alkaloids. The weight of this precipitate multiplied by 5 (or, if grains have been taken, divided by 2) gives the amount of cinchonine, quinidine, and amorphous alkaloids. This weight, added to that of the quinine and cinchonidine from Step II., gives the total alkaloids.

Step IV. Separation of Quinine and Cinchonidine.—This is an operation to effect which with absolute accuracy requires special experience, and to give the detailed instructions and solubility allowances by which alone it can be carried out to within tenths of a per cent. would be beyond the scope of the present work. It may, however, be generally stated that it is accomplished by dissolving the precipitate from Step II. in a little water acidulated with hydrochloric acid, then adding excess of sodium hydrate and shaking up with ether. After standing for some hours the ethereal layer is separated and evaporated to dryness. The residual quinine (now containing only a little cinchonidine) is dissolved in a fixed quantity of alcohol by the aid of a slight excess of dilute sulphuric acid, the alcohol is evaporated off, the residue is dissolved in a fixed quantity of water and brought to the boil. While hot, this solution is rendered as nearly neutral as possible by dropping in dilute ammonium hydrate (but not so as to produce a permanent precipitate) and set aside for twelve hours, when crystals of sulphate of quinine $(C_{20}H_{24}N_2O_2) \cdot H_2SO_4 \cdot 15H_2O$ deposit, leaving sulphate of cinchonidine in solution. These crystals are then collected and weighed with certain special precautions. The use of animal charcoal to perfectly decolorise is also necessary.

Step V. Separation of the quinidine, cinchonine and amorphous alkaloid. The precipitate from Step III. is digested in cold proof spirit, which dissolves the quinidine and amorphous alkaloids, and leaves the cinchonine insoluble for collection, drying and weighing. The alcoholic solution is rendered acid with acetic acid and evaporated to dryness on the water bath. The residue is dissolved in a very small quantity of water, and a little spirit and some solution of sodium iodide is added, which precipitates the quinidine as iodide. This is weighed and the weight multiplied by ·718 = quinidine. The amorphous alkaloids are then found by difference.

3. Estimation of Morphia in Opium.

Dry the opium in the water oven, powder it, and once more dry it at 212° till the weight is constant. Take 14 grammes (or 140 grains) of this powder, and 6 grammes (or 60 grains) of fresh slaked lime, and triturate them together with 40 c.c. (or 400 grain-measures) of distilled water in a mortar until a uniform mixture results; then add 100 c.c. (or 1000 grain-measures) of distilled water and stir occasionally during half an hour. Filter the mixture through a plaited filter about three inches in diameter in a wide-mouthed bottle or stoppered flask, having the capacity of about six fluid ounces and marked at exactly 104 c.c. (or 1040 grain-measures), until the filtrate reaches this mark. To the filtered liquid, representing 10 grammes (or 100 grains) of opium, add 11 c.c. (or 110 grain-measures) of rectified spirit, and 50 c.c. (or 500 grain-measures) of ether, and shake the mixture; then add 4 grammes (or 40 grains) of ammonium chloride, shake well and frequently during half an hour, and set it aside for twelve hours. Counter-balance two small filters; place one within the other in a small funnel, and decant the ethereal layer as completely as practicable upon the inner filter. Add 20 c.c. (or 200 grain-measures) of ether to the contents of the bottle, and rotate it; again decant the ethereal layer upon the filter, and afterwards wash the latter with 10 c.c. (or 100 grain-measures) of ether added slowly and in portions. Now let the filter dry in the air, and pour upon it the liquid in the bottle in portions, in such a way as to transfer the greater portion of the crystals to the filter. When the fluid has passed through the filter, wash the bottle and transfer the remaining crystals to the filter, with several small portions of distilled water, using not much more than 20 c.c. (or 200 grain-measures) in all, and distributing the portions evenly upon the filter.* Allow the filter to drain, and dry it, first by pressing between sheets of bibulous paper, and afterwards at a temperature between 131° and 140° F. (55° and 60° C.), and finally at 194° to 212° F. (96° to 100° C.) Weigh the crystals in the inner filter, counterbalancing by the outer filter. The crystals should weigh 1 gramme (or 10 grains), corresponding to 10 per cent. of morphine in the dry powdered opium. It is considered that anything between 9·5 and 10·5 per cent. is correct for official opium.

* When it is desired to obtain the morphine in a state of absolute purity, the precipitate should be washed first with water saturated with morphia, and then with rectified spirit similarly saturated.

IV. Estimation of the Alkaloidal Strength of Extracts.

Dissolve 1 gramme (or 10 grains) of the extract in 20 c.c. (or half a fluid ounce) of water, heating gently if necessary, and add 6 grammes (or 1 drachm) of sodium carbonate previously dissolved in 20 c.c. (or half a fluid ounce) of water and 20 c.c. (or half a fluid ounce of chloroform; agitate, warm gently, and separate the chloroform. Add to this 20 c.c. (or half a fluid ounce) of dilute sulphuric acid with an equal bulk of water; again agitate, warm, and separate the acid liquor from the chloroform. To this acid liquor add now an excess of ammonia, and agitate with 20 c.c. (or half a fluid ounce) of chloroform; when the liquors have separated, transfer the chloroform to a weighed dish, and evaporate the chloroform over a water bath. Dry the residue for one hour at 212° F., and weigh. Thus treated, extract of *nux vomica* should show 15 per cent of total alkaloids, and the process may be extended to almost any extract containing alkaloids except opium. No standards have, however, yet been fixed, except that for *nux vomica.*

V. Examination of a Tincture or other Alcoholic Liquor for the Presence of Methylated Spirit.

For the purpose of testing tinctures or any alcoholic liquors, they must first be distilled until a part of the spirit has passed over. The distillate is treated as follows:—

A small flask is fitted with a cork and a tube having two right-angular bends, with the end dipping into a test-tube kept cold by immersion in water; and in it is put,—

(1) About half a drachm of the spirituous liquid required to be tested.
(2) An equal quantity of potassium dichromate and of pure sulphuric acid.
(3) Four or five times as much water.

The mixture, after standing for twenty minutes, is distilled at a gentle heat, until nearly the whole has passed over. Sodium carbonate having been added to the distillate till it is slightly alkaline, the liquid is evaporated in a porcelain basin to about half its bulk, and having been acidulated slightly by acetic acid, is transferred to a test-tube, heated gently with twenty drops of a 5 per cent. solution of argentic nitrate for a few minutes, when any decided opacity (due to the discoloration of the fluid and the separation of a blackish precipitate of metallic silver), indicates the presence of methyl hydrate in the sample thus tested. In the oxidation of ordinary alcohol a mere trace of formic acid is formed by secondary decomposition, consequently a distinct precipitate must be obtained before the spirit can, with certainty, be pronounced to be methylated.

The operation of the above process depends on the fact that by a short oxidation with sulphuric acid and potassium dichromate, aldehyds and acids are produced, which, by being boiled with sodium carbonate, yield sodium acetate and formate, the former from the ordinary "alcohol" and the latter from the wood spirit. Upon the addition of argentic nitrate, argentic formate is produced, which is easily reduced by boiling to metallic silver, while argentic acetate is not so affected. If the oxidation be too powerful, the formic acid in turn becomes oxidised to CO_2 and H_2O and lost. Thus the process is not always satisfactory, and must be completed within half an hour or so.

When this process has to be applied to sweet spirit of nitre, the ethyl nitrite must be first got rid of, as follows :—

Take a little of the spirit and place it in a bottle with some dry potassium carbonate, and shake up. Let it settle, and take about two drachms of the strong spirit which separates. Saturate this with calcium chloride, and distil on a water bath, rejecting the distillate (ethyl nitrite, etc.). Add a little water to the contents of the retort and distil again, when the pure spirit will come over, and a portion may then be tested as above directed.

VI. Estimation of the Strength of Resinous Drugs.

Take 5 to 10 grammes of the drug in powder, and place it in a strong glass flask with 100 c.c. of pure methylated spirit (60° O.P. and free from resin). Close the flask with a good cork, and digest it in a warm place at about 120° F. for 12 hours, shaking from time to time. Pour or filter off 80 c.c. (representing $\frac{8}{10}$ of the total drug taken), place it in a weighed beaker, and evaporate to 25 c.c. on the top of the water bath. Now add 50 c.c. of distilled water and boil gently over a low gas flame till all the alcohol is driven off. Let it cool and perfectly settle, pour off the supernatant liquor, wash the deposited resin by decantation with hot distilled water, and then dry the beaker and its contents in the air bath at 220° F. and weigh, deducting the tare of the beaker. Thus treated, jalap, for example, should show 10 per cent. of resin, but for other resinous drugs no official standard has yet been laid down.

VII. Testing the Purity of Quinine Sulphate. (*Official Directions.*)

Test for Cinchonidine and Cinchonine.—Heat 100 grains of the sulphate of quinine in five or six ounces of boiling water, with three or four drops of diluted sulphuric acid. Set the solution aside until cold. Separate, by filtration, the purified sulphate of quinine which has crystallised out. To the filtrate, which should nearly fill a bottle or flask, add ether, shaking occasionally, until a distinct layer of ether remains undissolved. Add ammonia in very slight excess, and shake thoroughly, so that the quinine at first precipitated shall be redissolved. Set aside for some hours, or during a night. Remove the supernatant clear ethereal fluid, which should occupy the neck of the vessel, by a pipette. Wash the residual aqueous fluid and any separated crystals of alkaloid with a very little more ether, once or twice. Collect the separated alkaloid on a tared filter, wash it with a little ether, dry at 212° F. (100° C.), and weigh. Four parts of such alkaloid correspond to five parts of crystallised sulphate of cinchonidine or of sulphate of cinchonine.

Test for Quinidine.—Recrystallise fifty grains of the original sulphate of quinine, as described in the previous paragraph. To the filtrate add solution of iodide of potassium, and a little spirit of wine to prevent the precipitation of amorphous hydriodates. Collect any separated hydriodate of quinidine, wash with a little water, dry, and weigh. The weight represents about an equal weight of crystallised sulphate of quinidine.

Test for Cupreine.—Shake the recrystallised sulphate of quinine, obtained in testing the original sulphate of quinine for cinchonidine and cinchonine, with one fluid ounce of ether and a quarter of an ounce of solution of ammonia, and to this ethereal solution, separated, add the ethereal fluid and washings also obtained in testing the original sulphate for the two alkaloids just mentioned. Shake this ethereal liquor with a quarter of a fluid ounce of a 10 per cent. solution of caustic soda, adding water if any solid matter separates. Remove the ethereal solution. Wash the aqueous solution with more ether, and remove the ethereal washings. Add diluted sulphuric acid to the aqueous fluid heated to boiling, until the soda is exactly neutralised. When cold collect any cupreine sulphate that has crystallised out on a tared filter, dry and weigh.

Quinine sulphate should not contain more than 5 per cent. of other cinchona alkaloids.

VIII. Estimation of the Alkaloidal Strength of Scale Preparations.

Dissolve 5 grammes in 30 c.c. of water, place the solution in a separator, add 20 c.c. of chloroform, and then a slight excess of ammonium hydrate. Shake well and allow it to stand. When the chloroform has separated off clear, run it off into a small weighed flat-bottomed flask, and repeat the shaking successively with 10 c.c. and 5 c.c. of chloroform, always running it off into the same flask. Attach the flask by a cork to a condenser and distil off the chloroform, and then dry the residue for half an hour in the air bath at 220° F., and weigh. Thus treated, citrate of iron and quinine should show 15 per cent. of alkaloidal residue, which should be soluble in ether.

IX. The Estimation of Phenol.

The following is the method for determining the phenol quantitatively in crude carbolic acid:—20 c.c. of potassium hydrate solution (sp. gr. 1·25—1·30) are added to 20 c.c. of the crude carbolic acid. The whole is well shaken up, and, after half an hour, the mixture is made up to $\frac{1}{4}$ litre by the addition of water. The tarry constituents of the carbolic acid separate out and are removed by filtration. The residue is washed with lukewarm water till the wash-water is no longer alkaline. The whole filtrate is then treated with hydrochloric acid till faintly acid (this point is also indicated by the liquid changing color and turning brown), and made up to 3 litres. The small quantity of tarry matter left in the filtrate does not interfere in the titration which follows. The dilution is necessary, for, in titrating, the carbolic acid solution must not contain more than ·1 gr. in 25 c.c. 50 c.c. are now taken, and 150 c.c. of a solution containing 2·040 grs. sodium bromate and 6·959 grs. sodium bromide to the litre are added, together with 5 c.c. of concentrated hydrochloric acid; bromine is evolved and tribromophenol precipitated. After 20 minutes, during which the mixture is shaken up frequently, 10 c.c. potassium iodide solution (125 gr. potassium iodide to the litre) are added; potassium bromide is formed with the excess of free bromine, and iodine liberated. After about 5 minutes (not longer) starch solution is added, and the free iodine titrated with a sodium thiosulphate solution ("hypo"), containing 9·763 grs. per litre. The experiment should be first done upon ·2 gramme of pure crystallised carbolic acid dissolved in 50 c.c. of water, and the number of c.c. of the "hypo" solution used should be noted. A blank experiment on 150 c.c. of the bromide-bromate solution, with 5 c.c. of HCl and 10 c.c. of the iodide solution, should also be titrated. By deducting the c.c. of "hypo" used in the first check from that in the second, the number of c.c. of "hypo" equivalent to ·2 gramme real carbolic acid is ascertained, and the equivalent value of each c.c. of "hypo" in carbolic acid is calculated. In every analysis the number of c.c. of "hypo" used is deducted from that employed in the blank experiment, and the difference is calculated to carbolic acid present in the sample.

X. Estimation of the Fatty Acids in Soap.

Two grammes of the soap, in fine shavings, are shaken up in a separator, with a slight excess of dilute sulphuric acid to liberate the acids. Ether is then added, and the fatty acids which have been liberated are dissolved up in it. When the decomposition of the soap is complete, the liquid below the ethereal solution is removed. With a little care this can be done very completely without any loss of the ethereal solution. The ether is then shaken up with distilled water, and the latter drawn off as before, and this process of washing repeated three times more. When all but a few drops of the wash-water have been drawn off, a few drops of barium chloride solution are added, the mixture shaken up, and the last traces of sulphuric acid thus removed. With a little practice so little water is left below the ethereal solution that the latter can be directly drawn off and evaporated in a weighed dish, and the residue finally dried in the water oven and weighed. The fatty acids obtained are perfectly free from sulphuric and hydrochloric acid, and do not get brown at 100° C.

XI. Estimation of Oleic Acid.

One gramme of the impure fatty acid is saponified in a basin by heating with a slight excess of alcoholic potash till dissolved, and then diluted with water. This solution is treated with acetic acid drop by drop, until on stirring a faint *permanent* turbidity ensues. Dilute solution of potassium hydrate is then stirred in drop by drop till the liquid just clears up, and then solution of plumbic acetate is stirred in until precipitation ceases. The precipitate having been allowed to settle, the supernatant liquor is poured off and the soap washed once with boiling water. A little clean sand is rubbed up with the soap in the basin, and the whole scraped out and transferred to a "Soxhlet," in which it is thoroughly exhausted with 90 c.c. of *pure* ether. The ethereal solution (which now contains only plumbic oleate, the plumbic palmitate and stearate being left insoluble in the Soxhlet) is transferred to a special apparatus, sold by apparatus vendors as "Muter's oleine tube." This is a graduated and stoppered tube holding 120 c.c., and having a spout and stopcock at 30 c.c. from its base. Previously to introducing the ether, place 20 c.c. of dilute hydrochloric acid (1 in 3) into the tube, and then make up the whole with ether rinsings of the basin to the 120 c.c. mark. Close the tube, shake well and set aside. When settled, note the full volume of the ethereal solution of oleic acid, and run off an aliquot part from the tap into a weighed dish, evaporate, dry in the water oven and weigh. Finally calculate this weight to that of the whole bulk of ethereal solution previously noted, thus getting the amount of real oleic acid present in the gramme of crude acid started with.

XII. Estimation of Glycerine.

This process depends on (1) obtaining the glycerine free from other oxidisable bodies, such as alcohol sugars, etc.; (2) oxidising it with potassium bichromate in the presence of sulphuric acid to carbon dioxide and water by the following equation:

$$3C_3H_5(OH)_3 + 7K_2Cr_2O_7 + 28H_2SO_4 = 14Cr_2(SO_4)_3 + 7K_2SO_4 + 9CO_2 + 40H_2O;$$

and (3) estimating the amount of CO_2 evolved.

The relation between carbon dioxide and glycerine is that 396 parts by weight of CO_2 represent 276 parts of glycerine, or in other words the weight in grammes of CO_2 found multiplied by ·69697 gives the amount of glycerine in the quantity weighed out for analysis. The process is conducted in an ordinary apparatus for the estimation of carbon dioxide like that figured on page 145, fig. 36. The fluid containing the glycerine is first tested for the presence of sugars, both cane and glucose, and if they are absent a measured quantity, (say 100 c.c.) is evaporated on the water bath with 3 c.c. of milk of lime and 2 grammes of pure sand, till a fairly detachable residue is obtained. This is then scraped from the basin and extracted in the "Soxhlet" with absolute alcohol of B.P. strength. The alcohol is distilled off in a weighed flask and the weight of the residue taken. The residue is then dissolved in a little distilled water, and such an aliquot part taken for analysis as shall not contain more than ·5 gramme of glycerine. This is introduced into the apparatus with 4 grammes of potassium bichromate in saturated solution, and into the acid reservoir are placed 10 c.c. of strong sulphuric acid, the absorbing tube being also charged with the same acid. The apparatus is weighed, and then the acid in the reservoir is allowed to mix with the glycerine and bichromate. Action sets in, and after standing for three hours in a warm place, the whole is heated to gentle ebullition, cooled and weighed. The loss of weight equals the CO_2 evolved, and this is then calculated to glycerine as above and lastly corrected from the aliquot part to the whole. In the presence of sugar the liquid must first be inverted by boiling with dilute hydrochloric acid (if cane sugar be present) and then evaporated with an excess of barium hydrate instead of lime. This will either decompose the sugars, or render them insoluble in the alcohol subsequently applied.

DIVISION II. MICRO-CHEMICAL IDENTIFICATION OF DRUGS.

Many of the active principles of drugs are best identified by first isolating them by suitable processes, and then dissolving in alcohol, ether, water or some other solvent, and allowing a drop or two of the liquid to crystallise on a microscope slide.

This is a subject of study possessing the utmost fascination, and with which familiarity is only acquired by practice.

Experience is necessary in selecting the most suitable solvent from which to crystallise an alkaloid, as the duration of the evaporation may have a marked effect upon the form of the crystals. In some cases evaporation may be accelerated by the aid of heat; in others, such a proceeding is fatal to success. The addition of alcohol to ether, and of water to alcohol appears to be the best means of retarding the process when necessary. To take the case of cocaine. From chloroform no crystals are deposited; from ether they are ill defined; but from alcohol, allowing evaporation to proceed very slowly, we get the best results. Always employ polarised light by which to view the crystals, either with or without the addition of a selenite plate. Here, again, the duration of evaporation has a marked effect, also the strength of the solution. If the substance is deposited in a thin film it may be altogether invisible without polarised light. Thick crystals frequently produce color without the selenite, and those that are very thick may depolarise without any coloration. This being borne in mind, no difficulty is experienced in practice, as it is easy to compare with an alkaloid of known purity crystallised under the same conditions. In the accompanying plates (drawn by Mr. A. Percy Smith, F.I.C.) are given representations of various substances crystallised under the best conditions, with the name of the solvent and the linear magnification. The letter B signifies a black field (ordinary polarised light) and V a violet field produced by the selenite.

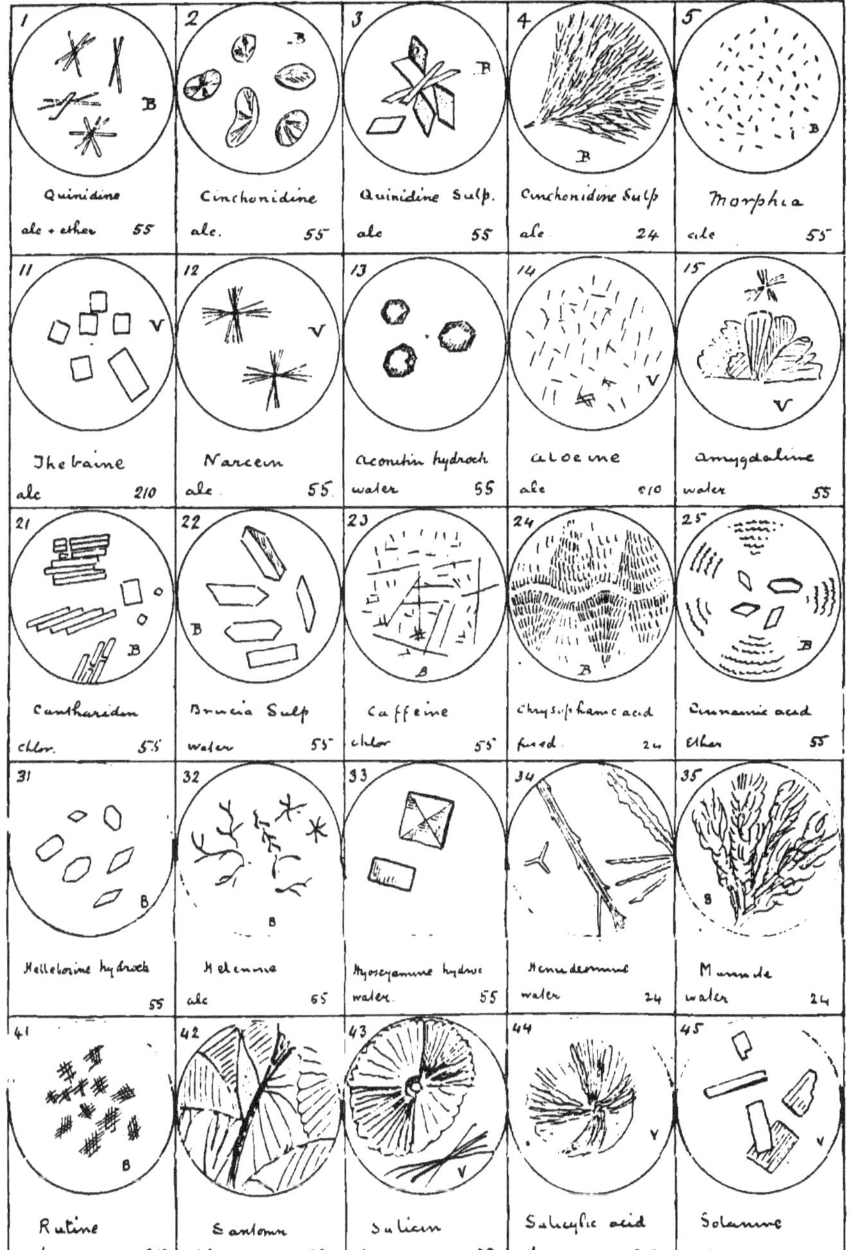
1
Quinidine
alc + ether 55
2
Cinchonidine
alc. 55
3
Quinidine Sulp.
alc 55
4
Cinchonidine Sulp
alc 24
5
Morphia
alc 55
11
Thebaine
alc 210
12
Narcein
alc. 55
13
aconitin hydroch
water 55
14
aloeine
alc 210
15
amygdaline
water 55
21
Canthariden
chlor. 55
22
Brucia Sulp
water 55
23
Caffeine
chlor 55
24
Chrysophanic acid
fused 24
25
Cinnamic acid
Ether 55
31
Helleborine hydroch
55
32
Helenine
alc 55
33
Hyoscyamine hydroc
water 55
34
Hemidesmine
water 24
35
water 24
41
Rutine
water 210
42
Santonin
chlor 55
43
Salicin
alc 55
44
Salicylic acid
ether 210
45
Solanine
water

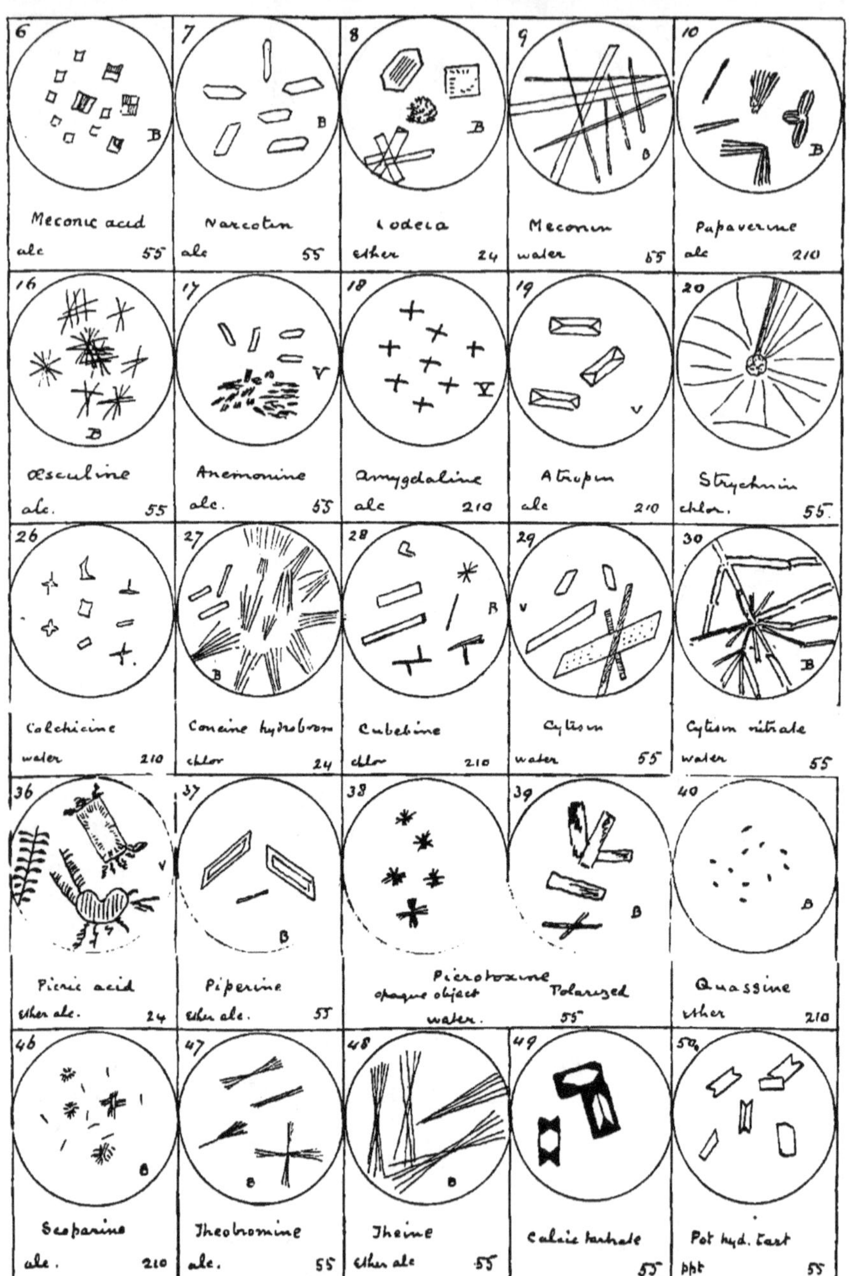

6
Meconic acid
alc 55
7
Narcotin
alc 55
8
Codeia
ether 24
9
Meconin
water 55
10
Papaverine
alc 210
16
Æsculine
alc. 55
17
Anemonine
alc. 55
18
Amygdaline
alc 210
19
Atropin
alc 210
20
Strychnin
chlor. 55
26
Colchicine
water 210
27
Conine hydrobrom
chlor 24
28
Cubebine
chlor 210
29
Cytisin
water 55
30
Cytisin nitrate
water 55
36
Picric acid
ether alc. 24
37
Piperine
ether alc. 55
38
39
Picrotoxine
opaque object
Polarized
water. 55
40
Quassine
ether 210
46
Scoparine
alc. 210
47
Theobromine
alc. 55
48
Theine
ether alc 55
49
Calcic tartrate
55
50
Pot hyd. tart
ppt 55

DIVISION III. ANALYSIS OF URINE.

The following are the chief points on which information is usually required by the physician who submits urine for examination to an analyst :—

1. Take the specific gravity, which should range from 1·015 to 1·025 at 60° F.

Note.—In diabetes the gravity is too high, sometimes reaching 1·060, while in albuminuria it is abnormally low, even occasionally falling to 1·005.

2. Examine the reaction, which should be very faintly acid.

3. Set a portion to settle in a long glass, and examine the deposit under the microscope for calcium oxalate or phosphate, uric acid or urates, pus, casts of kidney tubes, etc., etc.

Note.—The nature of the deposit may also be confirmed chemically as follows :—

(*a*) Warm the urine containing the sediment, when, if the latter should dissolve, it consists entirely of urates. In this case let it once more crystallise out, and examine it by the ordinary course for Ca, Na, and NH_4, to ascertain the bases.

(*b*) If the deposit be not dissolved by heating, let it settle, wash once by decantation with *cold* water, and warm with acetic acid. Phosphates will dissolve, and may be reprecipitated from the solution by excess of NH_4HO filtered out, well washed with boiling H_2O, dissolved in $HC_2H_3O_2$ and examined for Ca or Mg by the usual course for these metals in presence of PO_4.

(*c*) If the deposit be insoluble in acetic acid, warm it with HCl. Any soluble portion is calcium oxalate, which may be precipitated by NH_4HO.

(*d*) If the deposit be insoluble in HCl it is probably uric acid. In this case apply the *murexid* test as follows. Place it in a small white dish, remove moisture by means of a piece of bibulous paper, add a drop or two of strong HNO_3 and evaporate to dryness at a gentle heat. When cold add a drop of NH_4HO, which will produce a purple color deepened to violet by a drop of KHO.

4. Test for albumin, as follows :—

(*a*) *Boiling test.*—Filter the urine, place 10 c.c. in a narrow test-tube, and add one drop of acetic acid. Heat the tube over a small flame in such a way that the *upper* portion of the liquid only shall be heated. Coagulation will take place, and the presence of albumin will be evident from the formation of a turbidity ranging from a faint cloud to a dense coagulum, but always strongly contrasted with the clear liquid beneath, which was not heated. Mucin also precipitates with this test.

(*b*) *Nitric test.*—To five volumes of cold *saturated* solution of magnesium sulphate add one volume of nitric acid (sp. gr. 1·42), and preserve this reagent for use. Pour some perfectly clear filtered urine into a tube and carefully add an equal volume of the reagent delivered gently from a pipette so that the liquids shall not mix. An opalescent zone will form at the point of contact either immediately or within twenty minutes, according to the quantity of albumin present. This zone should not dissolve on gently warming, but should be a distinct ring at the bottom of the urine and not a general haze near the top, which latter indicates mucin.

(*c*) *Picric acid test.*—Dissolve 7·5 grammes of pure crystallised trinitrophenol (picric acid) in 500 c.c. of water, let it stand for some days to perfectly clarify, pour off and preserve the reagent for use. Mix some of the filtered urine in a tube with an equal volume of this reagent, look for any cloud or precipitate, and then heat to boiling. The true albumin cloud will remain permanent, while that due to peptones or alkaloids accidentally present will be dissolved. Picric acid does not precipitate mucin, and is therefore a valuable confirmation test.

(*d*) *Bödeker's method.*—Take a drachm of the urine, acidulate it with acetic acid, and add some potassium ferrocyanide drop by drop till a clear excess has been added. If during the addition a precipitate forms, albumin is to be suspected. Mere traces require some time to cause the cloud.

(*e*) *To estimate the Albumin.*—This may be done empirically by means of an *albuminimetre*, which can be purchased (with full instructions for use) at apparatus depôts. The action depends on observing the height of the albumin precipitate in a specially constructed graduated cylinder after standing for a given time. The precipitant is picric acid. In the absence of such a convenient appliance we may take a weighed quantity of the urine, and allow it to drop into boiling water acidulated with acetic acid. Collect the precipitate on a tared filter, wash with boiling water, dry at 212°, weigh, and deduct the weight of the filter, when the balance=albumin in the weight of urine operated upon.

5. Test for grape sugar, as follows:—

(*a*) *Moore's test.*—Acidulate with acetic acid, boil, and filter out any albumin if necessary. Then mix the filtrate with equal parts of *liquor potassæ* and heat to boiling, when ordinary urine will turn brownish-red, but saccharine urine will become dark brown or black.

(*b*) *Boettger's test* (modified by Nylander).—Dissolve 2·5 grammes of *pure* bismuth oxynitrate (free especially from silver) and 4 grammes of Rochelle salt in 100 grammes of 8 per cent. solution of sodium hydrate, and preserve for use. To use this reagent 1 c.c. of urine is added to 10 c.c. and the whole boiled gently for some time, when if even only traces of sugar be present the mixture becomes black.

(*c*) *Fehling's test.*—Render the urine alkaline with potassium hydrate, and filter to remove any phosphates, etc., which may precipitate, Boil the filtrate with Fehling's solution of copper (see page 119). and if a red precipitate should form, sugar is present.

(*d*) *To estimate the sugar.*—This is best done by taking 10 grammes of the urine and diluting it with water to 100 c.c. Place this solution into a burette, and run it gradually into 10 c.c. of Pavy's solution, kept boiling in a flask as directed under the Volumetric Analysis of Sugar, page 119. The number of c.c. of urine used will contain ·005 gramme of grape sugar, and then $\frac{100 \times \cdot 005}{\text{c.c. used}}$ = sugar in the 10 grammes of urine taken.

6. Test for bile, as follows:—

(*a*) *Oliver's test.*—Dissolve 2 grammes of flesh peptone, ·25 gramme of salicylic acid, and 2 c.c. of 33 per cent. acetic acid, in enough water to yield 200 c.c. of product. The solution should be rendered perfectly brilliant by passing it through frozen filtering paper. The urine, which should be very clear, is diluted to a specific gravity of 1·008. One cubic centimetre of this is added to 3 c.c. of Oliver's reagent. An opalescence at once appears, which will be found to be more or less distinct according to the quantity of bile salts present. Keller's contact-method can be advantageously employed for applying the test.

(*b*) *Gmelin's test for bile pigments.*—Place a drachm of nitric acid in a test-tube and cautiously pour upon it an equal volume of the urine. In the presence of bile a play of colors from green to violet, blue, and red will be observed where the liquids touch.

(*c*) *Pettenkofer's test for biliary acids.*—Mix equal parts of urine and sulphuric acid, add one drop of saturated syrup, and apply a gentle heat. If biliary acids be present, the color will change from cherry-red to deep purple.

Note.—Bilious urine is usually of a brownish-green color.

7. Test for urea, as follows:—

(*a*) Separate any albumin (as directed in Moore's test) if necessary, and evaporate an ounce of the urine to a syrupy consistence on the water bath. When cold add nitric acid, drop by drop, till crystals of nitrate of urea cease to deposit.

(*b*) *Estimation of urea.*—This is performed by the hypobromite process already given at page 124.

8. Test for uric acid by mixing one ounce of the urine with one drachm of hydrochloric acid in a beaker, and set aside for some hours. The uric acid will be deposited in reddish-brown crystals, which may, if desired, be weighed and proved by the *murexid* test.

This method is based on the known fact that argentic urate, unlike most other salts of silver, is insoluble in ammonia, but dissolves in nitric acid. The solutions required are:—1. "$\frac{N}{100}$ ammonium-thiocyanate"; dissolve about 8 grammes of ammonium-thiocyanate in a litre of water, and check with $\frac{N}{10}$ argentic-nitrate solution; dilute it for use with nine volumes of water. 2. Dissolve 5 grammes of argentic nitrate in 100 c.c. of distilled water, and add ammonia until the solution becomes clear. 3. Dilute 70 per cent. nitric acid with two volumes of distilled water, boil, to destroy the lower oxides o nitrogen, and preserve from the action of light. 4. A saturated solution of ferric alum. 5. Strong solution of ammonia. The following is a descript on of the process:—Place 25 c.c. of urine in a beaker with 1 gramme of sodium bicarbonate. Add 2 or 3 c.c. of strong ammonia, and then 1 or 2 c.c. (or an excess) of the ammoniated silver solution. A special procedure is necessary in order to collect the precipitate, as follows:—Fill a glass funnel to about one-third with broken glass, and cover this with a bed of good asbestos to about a quarter of an inch deep. This is best done by shaking the latter in a flask with water until the fibres are thoroughly separated, and then pouring the emulsion so prepared in separate portions on to the broken glass. On account of the nature of the precipitate and of the filter, it is necessary to use a Bunsen water pump in order to suck the liquid through. Having thus collected the precipitate, wash it with distilled water until the filtrate ceases to become opalescent with a solution of NaCl. Now dissolve the precipitate by washing it through the filter into a beaker, with a few cubic centimetres of the special nitric acid. Estimate the silver by Volhard's method, thus: Add to the liquid in the beaker a few drops of the ferric alum solution to act as an indicator, and from a burette carefully drop in $\frac{N}{100}$ ammonium thiocyanate solution until a permanent red color appears. The number of c.c. used multiplied by 0·00168, gives the amount of uric acid in the 25 c.c. of urine. One milligramme may be added to this amount as an allowance for average loss, and the whole multiplied by 4 gives the percentage of uric acid in the urine. The sodium bicarbonate is added in the early part of the process, to prevent decomposition of the argentic urate, which would otherwise occur.

9. Test for phosphates, as follows :—

(*a*) Add to one ounce of the urine a slight excess of ammonium hydrate, and boil. $Ca_3(PO_4)_2$ and $MgNH_4PO_4$ will both be precipitated, and the precipitate, if more than a distinct cloud, should be filtered out, dissolved in HCl, and analysed by the ordinary process already given for phosphates.

(*b*) After filtering out the earthy phosphates as above, alkaline phosphates may be tested for by adding *magnesia mixture* to the filtrate and getting the usual precipitate of $MgNH_4PO_4$ after standing some hours in a cold place.

(*c*) *Estimation of Phosphates.*—This is done by the volumetric process with uranic nitrate, already described at page 120.

10. Test for sulphates, as follows :—

Acidulate a weighed quantity of the urine with HCl, warm, and add excess of $BaCl_2$. If the precipitate appear too copious, estimate as usual.

11. Test for chlorides, as follows :—

Acidulate a little of the urine with HNO_3 and add excess of argentic nitrate. If the precipitate thus produced looks very large, a weighed quantity of the urine should be taken, and the chlorides estimated by Volhard's method (see page 113).

12. *Blood* is best seen under the microscope; but urine containing it has always a very characteristic smoky appearance. A test for blood is to add tincture of guaiacum and ethereal solution of hydrogen peroxide, which produce a sapphire blue; but such color of itself should not be taken as positive proof without the blood discs being also visible under the microscope.

DIVISION III. ANALYSIS OF URINARY CALCULI.

The following table will show at a glance the compositions and methods of proving the various calculi.

1. Calculi, fragments of which, heated to redness on platinum, entirely burn away.

NAME.	PHYSICAL CHARACTERS.	CHEMICAL CHARACTERS.
Uric acid, $C_5N_4H_4O_3$	Brownish-red; smooth or tuberculated; concentric laminæ (common)	Insoluble in water; soluble in KHO by heat, but evolves no NH_3; dissolves with effervescence in HNO_3. and the residue on evaporating the solution is red and gives the *murexid* test.
Ammonium urate	Clay-colored; usually smooth, and rarely with fine concentric laminæ (uncommon)	Soluble in hot water; soluble in heated KHO, evolving NH_3. Behaves with HNO_3 like uric acid.
Cystine, $C_3H_7NSO_2$	Brownish-yellow, semi-transparent and crystalline (very uncommon)	Insoluble in H_2O, alcohol, and ether. Soluble in NH_4HO, and depositing, when allowed to evaporate spontaneously, hexagonal plates. When heated, gives off odour of CS_2.
Xanthin, $C_5H_4N_4O_2$	Pale polished brown surface (very uncommon)	Soluble in KHO; soluble in HNO_3 *without effervescence*, and the solution leaves on evaporation a deep-yellow residue.

2. Calculi, fragments of which, heated to redness on platinum, do not burn away.

NAME.	PHYSICAL CHARACTERS.	CHEMICAL CHARACTERS.
Calcium oxalate, *mulberry calculus*, CaC_2O_4	Deep brown, hard and rough; thick layers (common)	Insoluble in acetic acid, but soluble without effervescence in HCl; heated to redness, it is converted into $CaCO_3$ which dissolves with effervescence in acetic acid, and the solution gives a white precipitate with $(NH_4)_2C_2O_4$. Heated strongly before the blowpipe, CaO remains, which, when moistened, is alkaline to test-paper.
Tricalcium phosphate, *bone-earth calculus*, Ca_3PO_4	Pale brown, with regular laminæ (uncommon)	Infusible before the blowpipe, and residue, when moistened, is not alkaline. Soluble in HCl, and the solution gives a *gelatinous* precipitate with excess of NH_4HO.
Magnesium ammonium phosphate, *triple phosphate calculus*, $MgNH_4PO_4$	White, brittle, crystalline, with an uneven and not usually laminated surface (uncommon)	Fusible with difficulty before the blowpipe, evolving NH_3, and residue not alkaline. Soluble in HCl, and solution gives white *crystalline* precipitate with NH_4HO.
Mixed phosphates of Ca, Mg, and NH_4, *fusible calculus*	White, and rarely laminated	Readily fusible before the blowpipe. Soluble in acetic acid, and solution gives a white precipitate with $(NH_4)_2C_2O_4$, and the filtrate from that precipitate gives a white precipitate with excess of NH_4HO.

CHAPTER XII.

THE ANALYSIS OF GASES, POLARISATION, AND SPECTRUM ANALYSIS.

DIVISION I. THE ANALYSIS OF GASES.

THIS operation is conducted by measuring a volume of the mixed gas under definite conditions of temperature and pressure, then exposing it to the action of some substance having the power of absorbing some one constituent of the mixture and again measuring the gas left. By seeing that the inside of the measuring tube is always kept moist the question of tension of aqueous vapor is equalised all through the experiment, and as many absorbents as may be necessary are employed in turn. Many of the gas-measuring appliances are large, costly, and require to be kept in special rooms devoted to the purpose. Quite recently, however, an American chemist, Mr. Keiser, has devised a gas-measuring apparatus which is likely, in the author's opinion, to supersede all others for absorption analyses, because it is compact in form, may be easily carried about and used at any place, and yet is capable of measuring gas volumes with great accuracy. Long graduated glass tubes and graduated vessels of all kinds are discarded, and an instrument is constructed upon the principle of determining the volume of a gas from the weight of mercury which it displaces at a known temperature and pressure. From the weight of mercury displaced the volume of the gas can be determined with much greater accuracy than by a direct reading on a graduated glass eudiometer.

The accompanying cut shows the construction of the measuring apparatus and the absorption pipette. A is the measuring apparatus, B is the absorption pipette; *a* and *b* are glass bulbs of about 150 c.c. capacity. They are connected at the bottom by a glass tube of 1 m.m. bore, carrying the three-way stopcock *d*. The construction of the key of the stopcock is shown in the margin. One hole is drilled straight through the key, and by means of this the vessels *a* and *b* may be made to communicate. Another opening is drilled at right angles to the first, which communicates with an opening extending through the handle, but does not communicate with the first opening. By means of this, mercury contained in either *a* or *b* may be allowed to flow out through the handle *d* into a cup placed beneath. The bulb *b* is contracted at the top to an opening 20 m.m. in diameter. This is closed by a rubber stopper carrying a bent glass tube, to which is attached the rubber pump *e*. To a second glass tube passing through the stopper a short piece of rubber tubing with a pinchcock is attached. By means of the pump *e* air may be forced into or withdrawn from *b*, as one or the other end of the pump is attached to the glass tube. The bulb *a* terminates at the top in a narrow glass tube, to which is fused the three-way stopcock *c*. The construction of the key of this stopcock is also shown in the cut. By means of it the vessel *a* may be allowed to communicate with the outside air, or with the tube passing to the absorption pipette, or with the gauge *g*. The gauge *g* is a glass tube having a bore 1 m.m. in diameter, and bent as shown in the figure. By pouring a few drops of water into the open end of this tube a column of water several centimetres high in both limbs of the tube is obtained. This serves as a manometer, and enables the operator to know when the pressure of the gas equals the atmospheric pressure. To secure a uniform temperature, the bulbs *a* and *b* are surrounded by water contained in a glass vessel. This vessel for holding water is merely an inverted bottle of clear glass from which the bottom has been removed. The handle of the stopcock *d* passes through a rubber stopper in the neck of the bottle. A thermometer graduated to $\frac{1}{5}^{\circ}$ is placed in the water near the bulb *a*. The whole apparatus is supported upon a vertical wooden stand.

The absorption pipette B consists of two nearly spherical glass bulbs of about 300 c.c. capacity. They communicate at the bottom by means of a glass tube, 3 m.m. inside diameter. *c* is a two-way stopcock. The holes in the key are drilled at right angles, so that the tube which connects with the

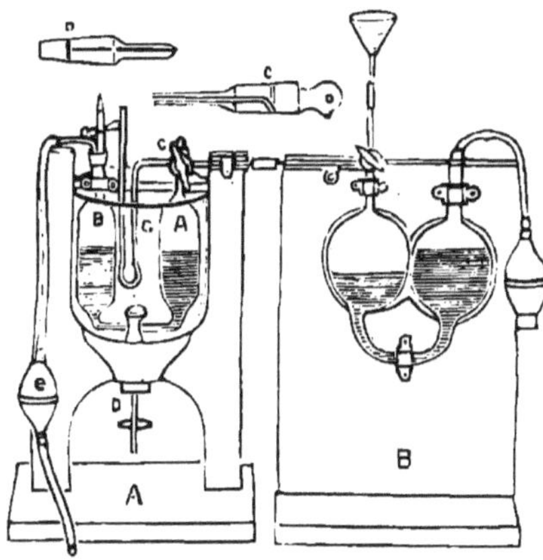

Fig. 46.

measuring apparatus may be put in communication either with the funnel or with the absorption bulb. The funnel is of service in removing air from the tube which connects the measuring apparatus with the absorption pipette. By pouring mercury or water into the funnel and turning the stopcocks *c'* and *c* in the proper directions all the air is readily removed. *f* is a rubber pump used in transferring gas from B to A. The lower part of the pipette contains mercury, which protects the reagent from the action of the air.

To measure the volume of a gas, the vessel *a* is filled completely with pure mercury. This is easily accomplished by pouring the mercury into *b*, and then, after turning *c* until *a* communicates with the outside air, forcing it into *a* by means of the pump *e*. Any excess of mercury in *b* is then allowed to flow out through the stopcock *d*. When *a* and *b* are now placed into communication the mercury will flow from *a* to *b*, and gas will be drawn in through the stopcock *c*. The volume of mercury which flows into *b* is equal to the volume of gas drawn into *a*. When the mercury no longer rises in *b* and it is desired to draw in still more gas into *a*, then it is only necessary to exhaust the air in *b* by means of the pump *e*. After the desired quantity of gas has been drawn into *a* the stopcock *c* is closed. After standing a few minutes the temperature of the gas becomes the same as that of the water surrounding *a*. The pressure of the gas is then made approximately equal to atmospheric pressure by allowing the mercury to flow out of *b* into a weighed beaker placed beneath the stopcock *d* until it stands at nearly the same level in both *a* and *b*. Communication is now established between *a* and *g*, and by means of the pump *e* the pressure can be adjusted with the utmost delicacy until it is exactly equal to atmospheric pressure. The stopcock *d* is then closed, and the remainder of the mercury in *b* is allowed to flow out into the beaker. The weight of the mercury displaced by the gas divided by the specific gravity of mercury at the observed temperature gives the volume of the gas in cubic centimetres.

If it is desired to remove any constituent of the gas by absorption, a pipette, B, containing the appropriate reagent, is attached to the measuring apparatus. All the air in the connecting tube is expelled by pouring mercury into the funnel and turning the stopcocks *c'* and *c* so that the mercury flows out through *c*. A little more than enough mercury to expel the gas in the vessel *a* is poured into *b*. The small quantity of air which is confined in the tube connecting *b* with the stopcock is removed by allowing a few drops of mercury to run out through *b*. Then *a* and *b* are placed in communication. The stopcocks *c'* and *c* are turned so that the gas may pass into the pipette, the mercury which filled the connecting tube passes into the absorbing reagent, and unites with that which is already at the bottom of the pipette. The transfer is facilitated by the pump *e*. After absorption the residual volume is measured in the same way that the original volume was measured. *a* is completely filled with mercury from the upper to the lower stopcock, and all the mercury in *b* is allowed to run out; the gas is then drawn back into the measuring apparatus, the last portion remaining in the connecting tube being displaced by means of mercury from the funnel. The volume is then determined as before.

The calculation of the results of an analysis is very simple. If the temperature and pressure remain the same during an analysis, as is frequently the case, then the weights of mercury obtained are in direct proportion to the gas volumes, and the percentage composition is at once obtained by a simple proportion.

If the temperature and pressure are different when the measurements are made, it is necessary to reduce the volumes to 0° and 760 m.m. The following formula is then used :—

$$V' = \frac{W(H-h)}{D(1 + \cdot 00367 \times t)\, 760},$$

in which

W = weight of mercury obtained (in grammes),
D = specific gravity of mercury at $t°$,
t = temperature at which the gas is measured,
H = height of the barometer,
h = tension of aqueous vapor,
V' = reduced gas volume (in cubic centimetres).

In all the measurements made with the apparatus the gas is saturated with aqueous vapor, because it comes in contact with the water in the manometer *g*.

The chief absorbents employed in gas analysis are as follows :—

A. Strong solution of potassium hydrate absorbs HCl, HBr, HI, CO_2, SO_2, and H_2S.

B. Crystallised sodium phosphate absorbs HCl, HBr, and HI.

C. 1 vol. of 25% solution of pyrogallic acid + 6 vols. 60% solution of KHO absorbs O_2 (after removal of any gas absorbed by KHO alone).

D. Concentrated solution of cuprous chloride in dilute hydrochloric acid absorbs CO (after removal of CO_2 and O_2 with alkaline pyrogallate).

E. A solution of sulphuric anhydride in strong sulphuric acid, or solution of bromine, absorbs C_2H_4, and the other gaseous hydrocarbons of the series C_nH_{2n}, and of C_nH_{2n-2}.

F. Absolute alcohol absorbs certain of the paraffins, except marsh gas.

G. Adding an excess of pure oxygen, and then absorbing with alkaline pyrogallate will remove NO together with the excess of oxygen used.

H. Hydrogen and nitrogen are left to be estimated by difference. They may be separated by passing the mixture into an eudiometer and adding excess of pure oxygen, measuring the total volume, and passing an electric spark. The hydrogen then forms water, and the gas being remeasured, $\frac{2}{3}$ of the total loss in volume represents the H_2 present. The excess of O_2 having then been removed by alkaline pyrogallate, the remainder is N_2.

Full details of the analysis of gases, beyond the scope of the present work, will be found in Sutton's "Volumetric Analysis."

DIVISION II. ANALYSIS BY CIRCULAR POLARISATION. THE SACCHARIMETER.

Crystals which do not belong to the regular system (notably calc-spar) possess the power of *double* refraction. That is to say, when a ray of light falls upon them, it is divided into two rays, one of which follows the ordinary rule of refraction, while the other takes a totally different course; and the two rays are called respectively the "**ordinary**" and the "**extraordinary**" ray. The most convenient polarising medium is what is called a "**Nicol's prism.**" It is composed of a crystal of calc-spar cut into two portions in the direction of its axis, and the two parts thus obtained cemented together with Canada balsam. When a beam of light enters the prism, it is doubly refracted by the *first* portion of the crystal, and the **extraordinary** ray only passes through the **second** portion to the eye of the observer; while the **ordinary** ray is completely reflected away by the layer of Canada balsam, and so lost to view. When this extraordinary ray is examined it is found to possess peculiar properties, such as showing color in transparent bodies which are usually colorless. This is accounted for by believing that it has become **polarised** —*i.e.*, that all its vibrations have been reduced to the same plane. If the polarised light thus obtained be examined by means of another Nicol's prism, it will be found that when the two prisms are placed with the principal sections parallel to each other, the ray will pass freely; but if the second prism, called the analyser, be then turned round so that its chief section is at right angles to that of the first, the polarised ray will in turn be entirely reflected from the layer of balsam, and no light will now reach the observer's eye. This holds good so long as nothing intervenes between the two prisms; but it has been found that certain bodies, such as quartz, possess the power, when interposed between the prisms, of giving a color instead of darkness, owing to their possessing the power of twisting the polarised ray from its original plane. Such substances are said to possess the power of **circular polarisation,** either in a "right-handed" or "left-handed" direction, according as it is necessary to turn the prism either to the right or left from its proper position to once more produce complete passage of the colorless polarised ray. The direction of the rotation is indicated by the use of arrows, thus: ↻ ↺. Cane sugar, grape sugar, dextrin, maltose, creasote, camphor, tartaric acid, cinchonine, castor oil, croton oil and oil of lemons rotate the plane of the polarised ray to the right; while fruit, or invert-sugar, quinine, cinchonidine, turpentine, and many essential oils, morphine, etc., have a left-handed rotation.

There are two varieties of quartz, known as right-handed and left-handed, one of which rotates the plane of polarisation to the right and the other to the left. If a plate of quartz 1 millimetre thick be placed between the two "Nicols," the ray of polarised light is rotated, and instead of being colorless, is colored, changing to all the colors of the spectrum as the analyser is turned, until it once more becomes colorless, and the amount that the analyser has to be turned (registered by a pointer on the degrees of the circle) is the index of rotary polarisation possessed by the quartz either in a right- or left-handed direction. If the turning of the analyser be now continued, color will again show itself, but this time it will be the color *complementary* to that at first produced. Thus, if we start with a plate of quartz showing red between the uncrossed prisms, and rotate, we shall find that when we have turned through an angle of 45°, we get no color, but after that we begin to get the complementary color green, which becomes most intense at the right angle of 90°, when the prisms are crossed. The polariscope as used for analysis is therefore essentially (*a*) a Nicol's prism acting as a polariser, (*b*) a plate of quartz usually divided down the centre, the one side being right-handed and the other left, (*c*) a tube to contain the solution, (*d*) another "Nicol" capable of being rotated, and having a pointer acting on degrees of the circle on a scale, (*e*) a telescope to focus the line between the two sides of the quartz. When the pointer is placed at zero, the tube filled with water and the line focussed, no color is seen on either side of it, but if a solution say of sugar be introduced, then color appears on one side of the line according to the nature of the sugar, and then the distance through which the pointer has to be moved round the graduated circle to get both sides of the quartz colorless is the degree of rotary polarisation. In practice monochromatic light from a sodium flame is employed, and this, destroying all color, causes a dark shadow instead of a color to appear when the instrument is used, so enabling color-blind persons to employ it without difficulty. To use the instrument we make a solution of the body to be examined of a definite percentage strength by dissolving a certain number of *grammes* in 100 c.c. of a solvent. We then fill the tube, observe the degree of rotation produced, and from that we calculate the absolute angle of rotation for the sodium light (always expressed as $[a]$) as follows:—

Let a=the observed angle, c the strength in grammes per 100 c.c., and l the length of the tube used in decimetres; then—

$$[a]_D = \frac{100a}{c \times l}$$

If the absolute angle thus found coincides with that obtained from the same substance in a state of purity, then the article under examination is pure, but if not, then a simple percentage calculation gives the impurity.

Thus the $[a]_D$ of pure cane sugar = . A sample examined as above gave an $[a]_D$ = .

Then: $\frac{. \times 100}{.}$ = per cent. of real sugar present in the sample.

DIVISION III. SPECTRUM ANALYSIS.

When a ray of sunlight is allowed to pass through a prism, it is deflected and *dispersed* into a number of rays differing in their degree of refrangibility. When these rays, as they pass from the prism, are caused to fall upon a white surface, they are observed to have a marked difference in color. The image so produced is called a **spectrum**; and when sunlight is thus treated it is found to give a spectrum consisting of the following colors: viz., violet, indigo, blue, green, yellow, orange, and red. The violet end of the spectrum, owing to its greater refrangibility, is always the nearer to the base or broad end of the prism. By this means of separating the rays of light we are able to ascertain the peculiar properties of each of the colors which go to compose it, and we find that the chemical activity of light resides chiefly in the most highly refrangible rays just outside the violet end of the visible spectrum, which are called the actinic rays; while, on the other hand, the heat transmitted by the sun is most felt at the opposite or red end of the spectrum.

Further research demonstrated that if we substituted the light emitted from various bodies in a state of incandescence to the action of a prism, the image or spectrum produced varied in each case, and was, moreover, almost characteristic of the particular bodies employed. This discovery led to the invention of the spectroscope, which, in its simplest form, consists of a metallic diaphragm with a narrow slit, through which a ray of light from the burning body is allowed to pass and is condensed by a lens upon a prism of glass, or, better still, a triangular bottle of thin glass filled with disulphide of carbon. At the opposite side of the prism is a short telescope, so arranged that an observer, looking through it, sees the spectrum or image produced by the light after passing through the prism. This telescope works upon a graduated scale, by which its position for viewing any particular line observed can be noted.

When ordinary solar light is examined through the spectroscope, a number of dark lines are found crossing the image at certain fixed points. They are called "**Frauenhofer's lines,**" and their position is characteristic of sunlight. It has been proved that such lines are only formed when the source of light contains volatile substances, as we find that the light emitted by a non-volatile heated body gives a continuous image devoid of lines. If, for example, a platinum wire be heated to a high temperature in a Bunsen burner, and the light thus produced be examined, no lines will be visible; but if the wire be now tipped with a fragment of sodium chloride, and once more ignited, a bright line will suddenly appear in the yellow of the spectrum, and in so dazzling a manner as to render the whole of the rest of the image almost invisible. In carrying out this system of analysis, therefore, it is only necessary to procure a perfectly clean piece of platinum wire, with one end bent into the form of a loop, and place a Bunsen gas burner in such a position that the rays from anything heated in it will pass into the spectroscope. The wire is then to be moistened with a little hydrochloric acid, and having been dipped in the substance to be examined, is to be held in the hottest portion of the Bunsen flame, and its spectrum simultaneously observed through the spectroscope, noting carefully the color, number, and position on the scale of the bright lines produced. When thus examined, we find that potassium exhibits one bright line in the red, and one in the blue; lithium, one bright line in the yellow, and one more brilliant in the red; strontium, one blue, one orange, and six red lines; barium, a number of lines chiefly green and yellow; calcium, three distinct bright yellow lines, one within green, and some broad but indistinct ones in the orange and red; and lastly, sodium the single bright yellow line already mentioned.

The student must commence with the examination of pure salts, carefully noting for reference the position of the index of the telescope on the scale where each characteristic line is found. When it is desired to examine any mixture the telescope index is brought to the required position and the substance is examined: if the proper line is seen, then the element searched for is present; if not, it is absent. If we examine ordinary light which has been made to pass through solutions of various colored bodies, we obtain dark bands analogous to the lines of Frauenhofer. These are called **absorption spectra**, and are very useful in the detection of soluble coloring matters. A solution of blood, for example, shows characteristic bands in the green of the spectrum. All this is a matter of special study, and to go farther into it would be beyond the scope of this volume.

DIVISION IV. MELTING POINTS.

The accurate taking of the melting point is an important factor in testing the purity of many solid organic bodies, notably of fats and waxes. Many methods have been from time to time proposed, but the following will be found to be sufficiently good for all ordinary purposes, and is, moreover, the method officially adopted in the Pharmacopœia.

A piece of narrow glass tube is softened in the gas flame and drawn out, so as to give it a long thin end with a capillary bore. The fat or wax is melted, and a little of it is sucked up into this capillary tube and allowed to solidify therein. The tube is then tied to a delicate thermometer so that its capillary end (having the semi-opaque column of solidified fat inside) just rests against the bulb of the thermometer. Both are now supported perpendicularly in a beaker of water, and heat is gently applied to the same. The rise of the thermometer and the appearance of the tube of fat being both observed, the height of the former, at the moment when the column of fat becomes transparent, is noted as the melting point required. It is always useful to repeat the process three times and to take the average as the true melting point.

Students desiring to go more deeply into this subject are referred to Allen's "Commercial Organic Analysis."

INDEX.

A.

Acetates, 51.
Acetic, acid, 51.
Acids, Acetic, 51.
  Antimonic, 50.
  Arsenic, 48.
  Arsenious, 48, 115.
  Benzoic, 56.
  Boric, 39.
  Carbolic, 55, 57.
  Carbonic, 38.
  Chromic, 49.
  Citric, 54.
  Cyanic. 43.
  Cyanuric, 43.
  Formic, 50.
  Fulminic, 44.
  Gallic, 57.
  Hydriodic, 32.
  Hydrobromic, 30.
  Hydrochloric. 29.
  Hydrofluosilicic, 40.
  Hydrosulphuric, 34.
  Lactic, 52.
  Malic. 53, 55.
  Meconic, 55.
  Metaphosphoric, 46.
  Nitric, 41, 123.
  Nitrous, 40.
  Oleic. 52, 180.
  Orthophosphoric, 47.
  Oxalic, 52, 146.
  Pyrophosphoric, 45.
  Salicylic, 56, 126.
  Silicic, 39, 146.
  Stannic, 50.
  Stearic, 52.
  Succinic, 53.
  Sulphuric, 37.
  Sulphurous, 36, 115.
  Tannic, 57.
  Tartaric, 53, 146.
  Uric, 182.
  Valerianic, 51.
Acidulous radicals, Detection of, 29, 75.
  Gravimetric estimation of, 141.
Acidimetry, 112.
Air, sanitary analysis, 163.
Albuminoid ammonia, 158.
Alcohol, estimation of in spirits. etc., 167.
  Table of percentages, 168.
Alkaline carbonates, estimation, 109.
  hydrates, estimation, 108.
Alkalies, Organic salts of, estimation, 110.
Alkaloids, table of reactions, 94.
  Detection. 89.
  Estimation, 121.
  Tests for, 94.
  Strength of extracts, 177.
Alkaloidal strength of scale preparations, 179.
Aluminium, 20.
  Estimation, 138.
Ammonia, 125.
  Albuminoid, 158.
Ammonium, 27.
  Estimation, 139.
Analysis—
  Bread, 167.
  Butter, 163.
  Colorimetric, 124.
  Colored sweets, 169.
  Drugs, 172.
  Food, 164.
  Gases, Keiser's apparatus, 189.
  Gravimetric, 132.
  Milk, 162.
  Mineral (of water), 147.
  Mustard, 169.
  Nitrometer, by the, 122.
  Pepper, 166.
  Polariscopic, 192.
  Qualitative, 1.
  Quantitative, 95.
  Sanitary, of water, 155.
  " " air, 163.
  Spectrum, 194.
  Starch, 169.
  Ultimate organic, 150.
  Urine, 184.
  Urinary calculi, 188.
  Vinegar, 171.
  Volumetric, 105.
Analytical factors, use of, 131.
Antimonic acid, 50.
Antimony, 16.
  Estimation, 136.
Apparatus, 106, 150.
Arseniates, 48, 145.
Arsenic, 15.
  Estimation. 136, 155.
Arsenic acid, 48.
Arsenious acid, 48.
  Estimation, 115.
Arsenites, 48.
Ash filters, 129.
,, of organic bodies, 131.

B.

Barium, 24.
  Estimation, 134.
Barks, cinchona analysis, 174.
Benzoates, 56.
Benzoic acid, 56.
Bile (urine), 186.
Bismuth, 14.
  Estimation, 134.
Blood (urine), 187.
Bread, analysis, 167.
Borates, 39.
Boric acid, 39.
Bromates, 31.
Bromides, 30.
  Estimation, 141.
  Separation, 42.
Bromine, 30.
  Estimation, 116.
Butter, analysis, 163.

C.

Cadmium, 15.
  Estimation, 134.
Calcium, 25.
  Estimation, 138.
Calculi, urinary, analysis, 188.
Carbolates, 55.
Carbolic acid, 55, 57.
Carbon, 38.
  Estimation, 151.
Carbonates, 38, 145.
  Alkaline, 109.
  Soluble, 123.
Carbonic acid, 38.

Cerium, 20.
Estimation, 138.
Chemical processes, 1.
Chlorates, 30.
Chlorides, 29, 42.
Estimation, 141, 187.
With bromides, detection of, 31.
With iodides, detection of, 32.
Chlorine, 29.
Estimation, 116, 155.
Available, estimation, 116.
Chromates, 49.
Chromic acid, 49.
Chromium, 21.
Estimation, 138.
Cinchona, analysis, 174.
Circular polarisation, 192.
Citrates, 54.
Citric acid, 54.
Clark's process, 161.
Cobalt, 23.
Estimation, 136.
Coefficients for analysis, 127.
Colorimetric analysis, 124.
Copper, 14.
Standard solution, 119.
Estimation, 134.
Crystallisation, 5.
Cyanates, 43.
Cyanic acid, 43.
Cyanides, 43, 141.
Cyanogen, 43.
Cyanuric acid, 43.

D.

Decantation, 3.
Density, vapor, 102.
Detection and separation of acidulous radicals, 29, 75.
Detection of alkaloids, 89.
Bromine, hydrobromic acid and bromides, 30.
Bromides in presence of iodides, 32.
Chlorides in presence of bromides, 31.
Chlorine, hydrochloric acid, and chlorides, 29.
Cyanides in presence of ferro-and ferri-cyanides, 45.
Iodate in an iodide, 33.
Inorganic acids, 78.
Metals, 10.
Nitrite in presence of a nitrate, 41.
Organic acids, 80.
Phosphate in presence of calcium, barium, strontium, manganese, magnesium, 48.
In presence of iron, 48.
Soluble sulphide in presence of sulphite and sulphate, 36.
Diagrams, Smith's, 182, 183.
Drying precipitates, 129.
Drugs, analysis, 169,
,, Micro-chemical analysis, 181.
,, Resinous (strength of), 178.
Dumas' process, 153.

E.

Electrolysis, 6.
Ether, Nitrous (estimation), 123.
Ethylsulphates, 51.
Estimation of—
Alkaline hydrates, 108.
,, carbonates, 109.
Alkaloids, 121.
Aluminium, 138.
Ammonia (Nesslerising), 125.
Ammonium, 140.
Antimony, 136.
Arseniates, 145.
Arsenic, 136, 155.
Arsenious acid, 115.
Ash filters, 129.
,, of organic bodies, 131.
Barium, 138.
Bismuth, 134.
Bromine, 116.
Cadmium, 134.
Calcium, 138.
Carbonates, 145.
Carbon dioxide, 163.
Carbon and hydrogen, 151.
Chlorine, 155.
,, free, 116.
,, available, 116.
Chromium, 138.
Cobalt, 136.
Copper, 134.
Fatty acids in soap, 180.
Ferric and ferrous salts, 117.
Glycerine, 180.
Gold, 135.
Gravimetric (of metals), 132.
Hydrocyanic acid, 114.
Hydrogen peroxide, 124.
Iodine, Free, 115.
Iron, 137.
Lead, 110, 132.
Manganese, 137.
Magnesium, 139.
Mercury, 133.
Moisture, 131.
Nickel, 137.
Nitric acid in Nitrates, 123.
Nitrogen, 152.
Nitrous Ether, 123.
Oleic acid, 180.
Organic salts of alkalies, 110.
Oxalic acid, 146.
Estimation of—
Phenol, 175.
Phosphates, 143.
Phosphoric acid, 120.
Phosphorus, 155.
Platinum, 135.
Potassium, 139.
Salicylic acid, 126.
Silicic acid, 146.
Silver, 132.
Sodium, 140.
Soluble carbonates, 123.
,, haloid salts, 113.
Sugar and starch, 120.
Sulphur, 155.
Sulphurous acid, 115.
Tartaric acid, 146.
Thiosulphites, 115.
Tin, 135.
Urea in urine, 124.
Zinc, 137.
Evaporation, 5.
Extraction, 2.
Extracts, alkaloidal strength of, 177.

F.

Factors, analytical (use of), 131.
Fehling's solution, 119.
Ferricyanides, 45.
Ferro- from ferri-cyanides, separation, 45.
Ferrocyanides, 44.
Ferric and ferrous salts, estimation, 117.
Filters, preparation of, 128.
ash, estimation, 129.
Filtration, 3.
Fluorides, 29.
Food analysis, 164.
Formic acid and formates, 50.
Fulminic acid, 44.
Fusion, 4.

G.

Gallic acid, 57.
Gallic, tannic, and pyrogallic acid, 57.
Gases, sp. gr., 101.
Analysis of, 189.
Gaseous impurities (testing for), 163.
Glycerine, 176.
Gold, 17.
Estimation, 135.
Gravimetric estimation, 132.
Acidulous radicals, 141.
Metals, 132.
Gravity, specific, 97, 184.
Group reagents, 10.
Aluminium, 20.
Ammonium, 27.
Antimony, 16.
Arsenic, 15.
Barium, 24.
Bismuth, 14.
Cadmium, 15.
Calcium, 25.

Group reagents (*contd.*)
Cerium, 20.
Chromium, 21.
Cobalt, 23.
Copper, 14.
Gold, 17.
Iron, 18.
Lead, 12.
Lithium, 26.
Manganese, 21.
Magnesium, 26.
Mercuricum, 13.
Mercuriosum, 13.
Nickel, 23.
Platinum, 18.
Potassium, 27.
Silver, 11.
Sodium, 27.
Strontium, 25.
Tin, 17.
Zinc, 22.

H.
Haloid salts, 113.
Hardness, Clark's process, 161.
Hydrates, 33.
Alkaline, estimation, 108.
Sodium, 112.
Hydriodic acid, 32.
Hydrobromic acid, 30.
Hydrochloric acid, 29.
Hydrocyanic acid, 43, 114.
Hydrofluoric acid, 29.
Hydrofluosilicic acid, 40.
Hydrogen, estimation, 151.
Peroxide, estimation, 124.
Hydrosulphuric acid 34.
Hypobromides, 31.
Hypobromic acid, 30.
Hypochlorites, 30.
Hypochloric acid, 29.
Hypophosphites, 36.
Hyposulphites, 46.

I.
Igniting precipitates, 130.
Indicator, 105.
Iodate with iodide, detection of, 33.
Iodates, 33.
Iodides, 32, 42.
Estimation, 141.
Iodine, 32.
Estimation, 115.
Standard solution, 114.
Iron, 18.
Estimation, 137.

K.
Keiser's apparatus, 189.
Kjeldahl's process, 154.

L.
Lactates, 52.
Lactic acid, 52.
Lead, 12.
Estimation, 110, 132.
Liquids (sp. gr. of), 97.
Lithium, 26.
Lixiviation, 2.

M.
Magnesium, 26.
Estimation, 139.
Malates, 53, 55.
Malic acid, 53.
Manganates, 49.
Manganese, 21.
Estimation, 137.
Manures, 161.
Estimation of phosphates, 144.
Mayer's standard solution, 121.
Measuring and weighing, 95.
Meconates, 55.
Meconic acid, 55.
Melting points, 195.
Mercuricum, 13.
Mercurosum, 12.
Mercury, estimation, 133.
Metals, detection of, 10.
Gravimetric estimation, 132.
In complex mixtures, 65.
Present in a salt, 61.
Separation into groups, 10, 66.
Tables for detection, 62.
Metaphosphoric acid and salts, 46.
Method, Volhard's, 113.
Varrentrapp and Will, 153.
Pavy's, 119.
Methylated spirit, in tinctures, 177.
Micro-chemical analysis, 181.
Milk analysis, 164.
Mineral analysis in water, 147.
Moisture, estimation, 131.
Mustard analysis, 169.

N.
Nesslerising, 125.
Nickel, 23.
Estimation, 137.
Nitrate, 40.
Nitrate, argentic, 113.
Nitrates, 41, 43, 157.
Nitric acid, 41.
Nitric acid estimation, 123.
Nitrite with nitrate, detection of, 41.
Nitrites, 40, 142.
Nitrogen estimation, 152, 157.
Nitrometer, use of, 105.
Analysis by, 122.
Nitrous acid, 40.
Nitrous ether, estimation, 123.

O.
Official standards of strength, 111.
Oleates, 52.
Oleic acid, 52.
Opium analysis, 176.
Organic analysis, ultimate, 150.
Matter, 163.
Orthophosphoric acid, 47.
Oxalates, 52, 146.
Oxalic acid, 52, 146.
Standard solution, 108.
Oxides, 34.
Oxygen required to oxidise, 160.

P.
Pavy's method, 119.
Pepper analysis, 169.
Perchlorates, 30.
Periodates, 33.
Permanganates, 49.
Peroxide hydrogen, 124.
Phenol and Phenates, 55. 179.
Phosphates, detection of, 46, 48.
Estimation, 143. 187.
Phosphoric acid, 156.
Estimation, 120.
Phosphorous acid, 46.
Phosphorus estimation, 155.
Platinum, 18.
Estimation, 135.
Poisons in mixtures (testing for), 92.
Polarisation, analysis by, 192.
Potassium, 27.
Estimation, 135.
Bichromate, standard solution, 117.
Preparation of sulphuretted hydrogen, 9.
Preparation of filters, 128.
Precipitates, drying, etc.,, 129.
Precipitation, 2.
Process, Clark's, 160.
Dumas', 163.
Kjeldahl's, 154.
Processes, chemical, 1.
Special analytical, 84.
Clark's for hardness, 161.
Testing for poisons, 92.
Pyrogallic acid, 57.
Pyrology, 7.
Pyrophosphoric acid and salts, 46.

Q.
Qualitative analysis—
Processes, 1.
Detection of metals, 10.
Detection and separation of Acidulous Radicals, 29.
Detection of unknown salts, 58.
Detection of Alkaloids, and of "Scale" medicinal preparations, 89.
Quantitative analysis—
Weighing and measuring, 95.
Specific gravity, 97.

Quantitative analysis(*contd.*)—
Vapor density, 102.
Standard solutions, 105.
Separations, 146.
Quinine sulphate (testing) 174.

R.

Radicals, acidulous, 29, 75, 141.
Reactions of the alkaloids (tables), 94
Reagents, 10.
Group I., 11.
Group II. Div. A, 13.
Group II. Div. B, 15.
Group III. Div. A, 18.
Group III. Div. B, 21.
Group IV., 24.
Group V., 26.
Resinous drugs (strength of), 178.

S.

Saccharimeter, 192.
Salicylic acid, 56.
Estimation, 126.
Salts, detection of unknown, 58.
Estimation of ferrous and ferric, 117.
Estimation of soluble haloid, 113.
Used in pharmacopœia, 64.
Sanitary analysis of air, 163.
Sanitary analysis of water, 156.
Scale preparations, 89.
Alkaloidal strength of, 179.
Separation — Group metals, 120—128.
Arseniate from phosphate, 49.
Chlorates from chlorides, 30.
Chlorides, iodides, bromides from nitrates, 42.
Cyanides from chlorides, 44.
Ferro- from ferri-cyanides, 45.
Iodide from bromide and chloride, 33.
Metals into groups, 66.
Oxalates, tartrates, citrates, malates, 55.
Quantitative, 146.
Silica from all other acids, 40.
Sulphides, sulphites and sulphates, 38.
Silica, 40.
Silicates, 39.
Silicic acid, 39, 146.
Anhydride, separation of, 40.
Silver, 11.
Estimation, 132.
Smith's diagrams, 182, 183.
Soap, fatty acids in, 180.
Sodium, 23.
Estimation, 139.
Hydrate, standard solution, 112.
Soil, estimation of phosphates, 144.
Solutions, standard, 105.
Separate, (treatment of), 173.
For testing acidulous radicals, 77.
Solubility tables, 82.
Specific gravity, 97.
Practical application, 101.
Liquids, 97.
Of urine, 184.
Solid bodies, 99.
Gases, 101.
Spectrum analysis, 194.
Spirits, alcoholic strength, 167.
Table for percentages, 168.
Standard solutions, 105.
Argentic nitrate, 113.
Barium chloride, 121.
Copper, Fehling's, 119.
For alkaloids, Mayer's, 121.
Iodine, 114.
Oxalic acid, 108.
Phosphate, 161.
Potassium bichromate, 117.
Sodium hydrate, 112.
Thiosulphate, 115.
Standards of strength, 111.
Stannates, 50.
Stannic acid, 50.
Starch estimation, 169.
Stearates, 52.
Stearic acid, 52.
Strength of extracts (alkaloidal), 177.
Resinous drugs, 178.
Standards of, 111.
Strontium. 25.
Sublimation. 4.
Succinates, 53.
Succinic acid. 53.
Sugar estimation, 120.
In urine, 185.
Sulphates, 36, 38.
Estimation, 142, 187.
Sulphides, 34, 36, 38.
Estimation, 142.
Sulphides, sulphites, and sulphates, separation of, 38.
Sulphites, 36, 38.
Sulphocyanates, 44.
Sulphovinates, 51.
Sulphur, 34.
Estimation, 155.
Sulphuric acid, 37.
Sulphurous acid, 36.
Estimation, 115.
Sulphuretted hydrogen, preparation of, 9.
Sweets (colored), analysis, 169

T.

Tables—
Coefficients for analysis, 127.
Degrees of thermometer, 164.
Detection of the metal in a simple salt, the metals as in pharmacopœia, 64.
Detection of the metal in a solution containing one base only, 62.
Distinction between gallic, tannic, and pyrogallic acids, 57.
General reaction of alkaloids, 94.
Percentages of alcohol, 168.
Tannic acid, 57.
Tartaric acid, 53, 146.
Tartrates, 53.
Testing for poisons, 92.
Chief alkaloids, 94.
Gaseous impurities, 163.
Quinine sulphate, 178.
Thiocyanates, 44.
Thiosulphates, 36, 115.
Tin, 17.
Estimation, 135.
Tinctures, methyl spirit in, 177.
Toxicological analysis. 89.

U.

Ultimate organic analysis, 150.
Urea, estimation, 124, 186.
Uric acid, 186.
Urinary calculi, analysis, 188.
Urine, 124.
Analysis, 184.

V.

Valerianates, 51.
Valerianic acid, 51.
Vaporisation, 5.
Vapor, specific gravity, 101.
Vapors, density of, 102.
Varrentrapp and Will, 153.
Vinegar analysis, 171.
Sulphuric acid in, 171.
Volhard's method, 113.
Volumetric analysis, 105.
Coefficients for, 122.

W.

Washing precipitates, 129.
Water, 33.
Mineral analysis of, 147.
Sanitary analysis of, 156.
Weighing precipitates, 129.
Will & Varrentrapp, 153.
Wines, alcoholic strength, 167.

Zinc, 22.
Estimation, 137.

www.ingramcontent.com/pod-product-compliance
Lightning Source LLC
Chambersburg PA
CBHW021706210326
41599CB00013B/1544

* 9 7 8 3 7 4 2 8 6 2 7 8 5 *